CAMBRIDGE LIBRARY COLLECTION

Books of enduring scholarly value

Physical Sciences

From ancient times, humans have tried to understand the workings of the world around them. The roots of modern physical science go back to the very earliest mechanical devices such as levers and rollers, the mixing of paints and dyes, and the importance of the heavenly bodies in early religious observance and navigation. The physical sciences as we know them today began to emerge as independent academic subjects during the early modern period, in the work of Newton and other 'natural philosophers', and numerous sub-disciplines developed during the centuries that followed. This part of the Cambridge Library Collection is devoted to landmark publications in this area which will be of interest to historians of science concerned with individual scientists, particular discoveries, and advances in scientific method, or with the establishment and development of scientific institutions around the world.

Polarisation of Light

Before his untimely death from typhoid, William Spottiswoode (1825–83) had served as president of the London Mathematical Society, the British Association, and the Royal Society. In addition to publishing widely in mathematics and the experimental physical sciences, he restored the fortunes of his family printing firm, Eyre and Spottiswoode, the Queen's Printers. An enthusiast for the popularisation of science, he lectured to large audiences at the Royal Institution, the South Kensington College of Science, and at British Association meetings. He also gave scientific talks at the school set up for the employees of his family firm. This illustrated 1874 work is based on these talks, and provides an introduction to 'this beautiful branch of optics'. Spottiswoode covers methods of polarisation, and the contemporary theory accounting for these effects. He describes various experiments, and explains how polarisation causes patterns and colours to appear in light.

Cambridge University Press has long been a pioneer in the reissuing of out-of-print titles from its own backlist, producing digital reprints of books that are still sought after by scholars and students but could not be reprinted economically using traditional technology. The Cambridge Library Collection extends this activity to a wider range of books which are still of importance to researchers and professionals, either for the source material they contain, or as landmarks in the history of their academic discipline.

Drawing from the world-renowned collections in the Cambridge University Library and other partner libraries, and guided by the advice of experts in each subject area, Cambridge University Press is using state-of-the-art scanning machines in its own Printing House to capture the content of each book selected for inclusion. The files are processed to give a consistently clear, crisp image, and the books finished to the high quality standard for which the Press is recognised around the world. The latest print-on-demand technology ensures that the books will remain available indefinitely, and that orders for single or multiple copies can quickly be supplied.

The Cambridge Library Collection brings back to life books of enduring scholarly value (including out-of-copyright works originally issued by other publishers) across a wide range of disciplines in the humanities and social sciences and in science and technology.

Polarisation of Light

William Spottiswoode

CAMBRIDGE
UNIVERSITY PRESS

University Printing House, Cambridge, CB2 8BS, United Kingdom

Cambridge University Press is part of the University of Cambridge.
It furthers the University's mission by disseminating knowledge in the pursuit of education, learning and research at the highest international levels of excellence.

www.cambridge.org
Information on this title: www.cambridge.org/9781108076234

This edition first published 1874
This digitally printed version 2015

ISBN 978-1-108-07623-4 Paperback

The original edition of this book contains a number of colour plates, which have been reproduced in black and white. Colour versions of these images can be found online at www.cambridge.org/9781108076234

POLARISATION OF LIGHT

LONDON: PRINTED BY
SPOTTISWOODE AND CO., NEW-STREET SQUARE
AND PARLIAMENT STREET

PREFACE.

THE FOLLOWING PAGES contain the substance of lectures delivered at various times to my workpeople, and constitute a talk rather than a treatise on Polarised Light. If a perusal of them should induce some to read, and others to write, more fully on the subject, this provisional sketch will have served its purpose. For such utility as it may possess I have to thank Professors Tyndall, G. G. Stokes, and Maskelyne, Sir C. Wheatstone, and others whose suggestions I have more or less consciously adopted; and last, but not least, those of my audience who, by patient attention to the lectures, have encouraged me to pursue the study of this beautiful branch of Optics.

CONTENTS.

CHAPTER VII.

CHAPTER VIII.

CHAPTER IX.

Description of the Plate (*at end*).

Figs.	
1 and 2.	Iceland Spar.
3 ,, 4.	Aragonite.
5 ,, 6.	Titanite or Sphene.
7 ,, 8.	Cyanide of Platinum and Barytes.
9 ,, 10.	Quartz perpendicular to the Axis.
11 ,, 12.	Airy's Spirals—right and left-handed Quartz superposed.

POLARISATION OF LIGHT.

CHAPTER I.

METHODS OF POLARISATION.

LIGHT is said to be polarised when it presents certain peculiarities, hereafter to be described, which it is not generally found to possess. These peculiarities, although very varied in their manifestations, have one feature in common, viz. that they cannot be detected by the unassisted eye; consequently, special instrumental means are required for their investigation.

The origin and meaning of the term polarisation will be better understood when some of the phenomena have been witnessed or described than beforehand, and I therefore postpone, for the present, an explanation of it.

The subject of polarisation may be approached by either of two roads, the experimental or the theoretical. The theoretical method, which proceeds upon the principles of the Wave Theory of Light, is remarkably complete and explicit; so much so that it not only connects together many very diversified phenomena, but even, in some cases, has suggested actual prediction. But inasmuch as the theory without experimental facts would be little better than a study of harmony without practical music, it will be best to begin with experiment.

It was stated above that certain instrumental means were requisite for detecting polarisation. Now there are various processes, some occurring in the ordinary course of natural phenomena, others due to instrumental appliances, whereby a ray of light may be brought into the condition in question, or "polarised;" and it is a fact both curious in itself and important in its applications, that any one of these processes (not necessarily the same as that used for polarising) may be used also as a means of examining whether the ray be in that condition or not. This latter process is called "analysation." When two instruments, whether of the same or of different kinds, are used, they are called respectively the "polariser" and the "analyser;" and the two together are included under the general name of "polariscope."

The four principal processes by means of which a ray of light may be polarised are, reflexion, ordinary refraction, double refraction, and scattering by small particles. These methods will be considered in order; but before doing so, it will be convenient to describe the phenomena of polarisation as exhibited by some instrument tolerably simple in its action, and of easy manipulation. For such a purpose a plate of crystal called tourmalin will perhaps serve better than any other to begin with.

Tourmalin is a crystal of which there are several

varieties, differing only in colour. Very dark specimens generally answer the purpose well, excepting that it is difficult to cut them thin enough to transmit much light. Red, brown, or green specimens are usually employed; the blue are, for the most part, optically unsuitable. Some white, or nearly white, specimens are very good, and may be cut into thicker plates without loss of light.

If we take a plate of tourmalin cut parallel to a particular direction within the crystal called the optic axis (the nature and properties of which will be more particularly explained hereafter), and interpose it in the path of a beam of light at right angles to the direction of the beam, the only effect perceptible to the unassisted eye will be a slight colouring of the light after transmission, in consequence of the natural tint of the particular piece of crystal. But if we examine the transmitted beam by a second similar plate of tourmalin, placed parallel to the former, the following effects will be observed. When the two plates are similarly placed, *i.e.* as if they formed one and the same block of crystal, or, as it is technically expressed, with their optic axes parallel, we shall perceive only, as before, the colouring of the light due to the tints of the two plates. But if either of the plates be then turned round in its own plane, so as always to remain perpendicular to the beam, the light will be observed to fade gradually, until, when the moving plate has been turned through a right angle, the light becomes completely extinguished. If the turning be continued beyond the right angle, the light will begin to revive; and when a second right angle has been completed, the light will be as bright as at the outset. In Figs. 1 and 2, *a, b, c, d, e, f, g, h,* represent the two plates; in Fig. 1 the two plates are supposed to be in the first position; in Fig. 2 the plate *e, f, g, h,* has been turned through a right angle. Of the parts which overlap, the shading in Fig. 1 represents the deepened colour due to the double thickness of the crystal; in Fig. 2 it indicates the complete extinction of the light. The same alternation of brightness and extinction will continue for every right angle through which the moving plate is turned. Now it is to be

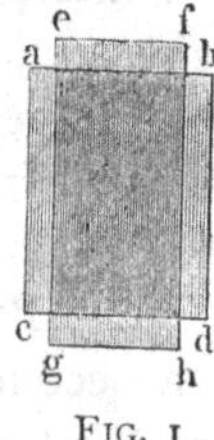

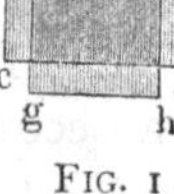

FIG. 1

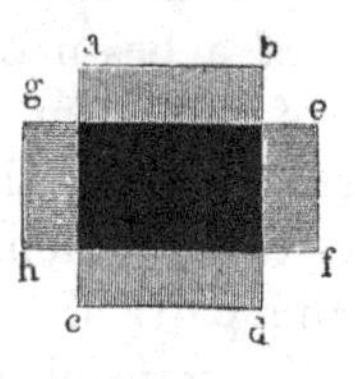

FIG. 2.

observed that this alternation depends only upon the angle through which one of the crystals has been turned, or, as it is usually stated, upon the relative angular position of the two crystals. Either of them may be turned, and in either direction, and the same sequence of effect will always be produced. But if the pair of plates be turned round bodily together, no change in the brightness of the light will be made. It follows, therefore, that a ray of ordinary light possesses the same properties all round; or, as it may be described in more technical language, a ray of ordinary light is symmetrical in respect of its properties about its own direction. On the other hand, a ray of light, after traversing a plate of tourmalin, has properties similar, it is true, on sides, or in directions, diametrically opposite to one another, but dissimilar on intermediate sides or directions. The properties in question vary, in fact, from one angular direction to another, and pass through all their phases, or an entire period, in every angle of 180°. This directional character of the properties of the ray, on account of its analogy (rather loose, perhaps) to the directional character of a magnet or an electric current, suggested the idea of polarity; and hence the condition in which the ray was found to be was called polarisation.

Having so far anticipated the regular order of things on the experimental side of the subject, it will perhaps be worth while to make a similar anticipation on the side of theory. It is considered as established that light is due to the vibrations of an elastic medium, which, in the absence of any better name, is called ether. The ether is understood to pervade all space and all matter, although its motions are affected in different ways by the molecules of the various media which it permeates. The vibrations producing the sensation of light take place in planes perpendicular to the direction of the ray. The paths or orbits of the various vibrating ethereal molecules may be of any form consistent with the mechanical constitution of the ether; but on the suppositions usually made, and none simpler have been suggested, the only forms possible are the straight line, the circle, and the ellipse. But in ordinary light the orbits at different points of the ray are not all similarly situated; and although there is reason to believe that in general the orbits of a considerable number of consecutive molecules may be similarly situated, yet in a finite portion of the ray there are a sufficient number of variations of situation to prevent any preponderance of average direction.

This being assumed, the process of polarisation is understood to be the bringing of all the orbits throughout the entire ray into similar positions. And in the case of the tourmalin plate the orbits are all reduced to straight lines, which consequently lie in one and the same plane. For this reason the polarisation produced by tourmalin, as well as by most other crystals, is called rectilinear, or, more commonly, plane polarisation. This property of tourmalin may also be expressed by saying that it permits only rectilinear vibrations parallel to a particular direction, determined by its own internal structure, to traverse it.

Adopting this view of polarisation as effected by a plate of tourmalin, it would be interesting to ascertain the exact direction of the vibrations.[1] And a simple experiment will go far to satisfy us on that point. The argument, as now stated at least, is perhaps based upon general considerations rather than upon strict mechanical proof; but the experimental evidence is so strong that it should not be denied a place here. Suppose for a moment that the tourmalin be so placed that the direction of vibration lies either in or perpendicular to the plane of incidence (that is, the plane

[1] It is still an open question whether the vibrations of polarised light are executed in, or perpendicular to, the plane of polarisation. Opinion inclines to the latter view, which is adopted throughout this work.

containing the incident ray, and a perpendicular to the surface on which it falls at the point of incidence); then it is natural to expect that vibrations executed in the plane of incidence will be far more affected by a change in the angle of incidence than those perpendicular to that plane. In fact, the angle between the direction of the vibrations and the surface upon which they impinge will, in the first case, vary with the angle of incidence; but in the second case it will remain unchanged.

In Figs. 3 and 4, *n o* represents the ray of light; the arrow the direction of vibration; *a, b, c, d, a′, b′, c′, d′*, the plate in two positions, turned in the first instance about the direction of vibration, in the second about a line perpendicular to it.

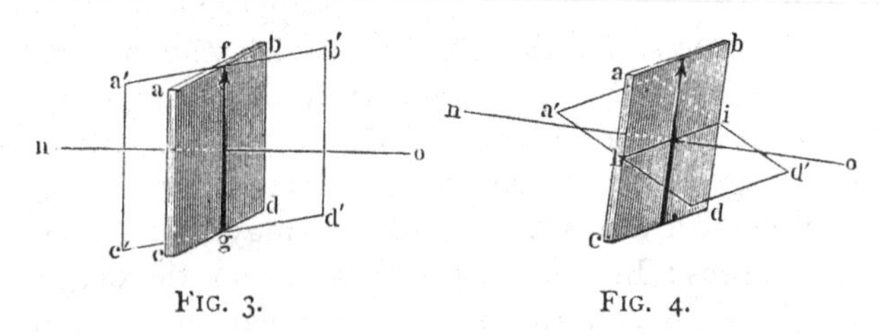

FIG. 3. FIG. 4.

Dismissing, then, the former supposition, and supposing that nothing whatever is known about the direction of vibration, then, if all possible directions be taken in succession as pivots about which to tilt or turn the second tourmalin, it will be found that for one direction the intensity of the light diminishes more rapidly with an increase of tilting (or, what is the same thing, with an increase of the angle of incidence) than for any other. And, further, that for a direction at right angles to the first, the intensity of light diminishes less than for any other; while for intermediate directions the diminution of intensity is intermediate to those above mentioned. In accordance, therefore, with what was said before, we may conclude that the vibrations are parallel to the line or pivot about which the plate was turned when the diminution of light was least.

Secondly, polarisation may be effected by reflexion. If light reflected from the surface of almost any, except metallic, bodies be examined with a plate of tourmalin, it will in general be found to show traces of polarisation; that is to say, if the plate of tourmalin be caused to revolve in its own plane, and the reflected rays be viewed through it, then in certain positions of the plate the reflected light will appear less bright than in others. If the angle at which the original rays fall upon the reflecting surface be varied, it will be found that the amount of alteration in brightness of the light seen through the revolving tourmalin (or analyser) will also vary. This fact may also be expressed thus: In polarisation by reflexion, the degree of polarisation, or the amount of polarised light in the reflected rays, varies with the angle of incidence on the reflecting surface. But at a particular angle, called on that account the polarising angle, the polarisation will be a maximum. This angle (usually measured between the incident ray and the perpendicular to the reflecting surface) is not the same for all substances; in fact, it varies with their refractive power according to a peculiar law, which, when stated in the technical language of science, may be thus enunciated: The tangent of the polarising angle is equal to the refractive index. Simple geometrical considerations, combined with the usual expressions for the laws of reflexion and refraction, will show that this relation between the polarising angle and the refractive index may be also expressed in the following way: If light be incident at the polarising angle, the reflected and refracted rays will be at right angles to one another.

In Fig. 5, *s i* represents the incident, *i f* the reflected, and *i r* the refracted ray. Then *s i* will be

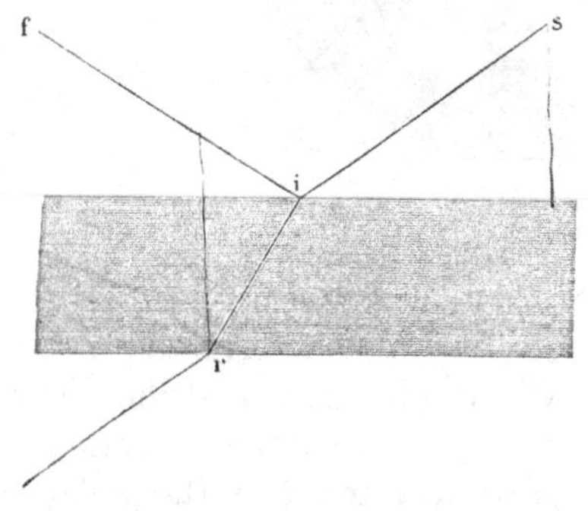

FIG. 5.

incident at the polarising angle when the angle *f i r* is a right angle.

An apparatus devised by Professor Tyndall for experimentally demonstrating the laws of reflexion and refraction is admirably adapted for verifying this law. The following description is quoted from his Lectures on Light: "A shallow circular vessel R I G (Fig. 6), with a glass face, half filled with water rendered barely turbid by the admixture of a little milk or the precipitation of a little mastic, is placed upon its edge, with its glass face vertical. By means of a small plane reflector M, and through a slit I in the hoop surrounding the vessel, a beam of light is

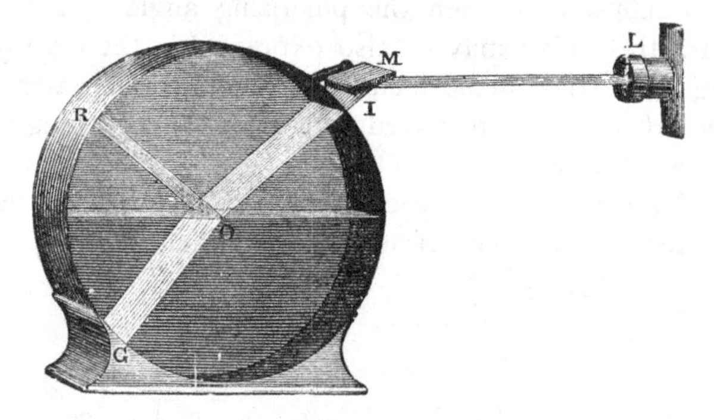

FIG. 6.

admitted in any required direction." If a little smoke be thrown into the space above the water, the paths of the incident, the reflected and the refracted beams will all be visible. If, then, the direction of the incident beam be so adjusted that the reflected and the refracted beams are at right angles to one another, and a Nicol's prism be interposed in the path of the incident beam, it will be found that, by bringing the vibrations alternately into and perpendicular to the plane of incidence, we shall render the intensity of the reflected and refracted rays alternately a minimum. Thus much for the verification of the law. But not only so; if we take different fluids, and for each of them in succession adjust the incident beam in the same manner, we shall only have to read off the angle of incidence in order to ascertain the polarising angle of the fluid under examination.

The polarising angle for glass is 54° 35′.

Returning to the general question of polarisation by reflexion, and calling to mind a principle stated above, viz. that any process which will serve for polarising will serve also for analysing, it follows that we may replace the analysing tourmalin by a second plate of glass (or whatever substance has been used for the first reflexion), placed parallel to the first, and in such a position as to receive the reflected ray.

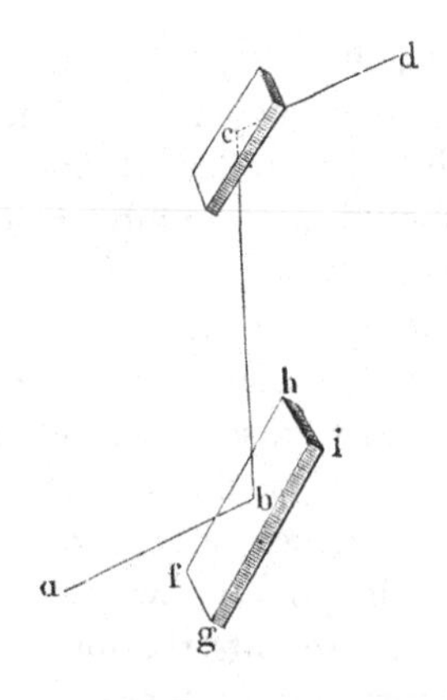

FIG. 7.

Thus, in Fig. 7, let *a b* be the incident, and *b c* the reflected ray at the first plate; *b c* the incident, and *c d* the reflected ray, at the second plate; then the ray *b c* will be polarised more or less, according to the angle of incidence, at *b*, and will be analysed at *c*. If the second plate be then turned round the ray *b c* reflected from the first plate as an axis, it will be found that at two positions of rotation (first, when the plates are parallel, and, secondly, when one of them has been turned through 180°) the light reflected from the second plate is brightest, and at two positions at right angles to the former the reflected ray is least bright. The degree of dimness at the two positions last mentioned will depend upon the accuracy with which the reflecting plates have been adjusted to the polarising angle; and when this has been completely effected, the light will be altogether extinguished.

Suppose, now, that the reflecting substance be, as in the case of glass, transparent. Then it will not be surprising if, when the reflected ray is polarised, the refracted ray should also exhibit traces of polarisation. And, in fact, every ray of ordinary light incident at any angle upon a transparent plate is partly reflected and partly refracted; the reflected ray is partially polarised, and so also is the refracted ray. This being so, if, instead of a single plate, we use a series of plates placed one behind the other, the light will be reflected in all possible ways from the two n surfaces, where n represents the number of the plates used. The phenomenon of these reflexions is, therefore, rather complicated; but the modifications due to the additional plates do not materially alter the proportion of polarised to unpolarised light. It is, however, otherwise with the refracted rays. The rays transmitted by the first plate enter the second in a state of partial polarisation, and by a second transmission undergo a further degree of polarisation. If this process be continued by having a sufficient number of plates, the ray finally emergent may have any degree of polarisation required.[1] And it is worthy of remark that, in proportion as the rays become more and more polarised, so does a less and less quantity of light become reflected from the surfaces of the plates; and, consequently, except in so far as light is absorbed by actual transmission through the substance of the plates, the emergent ray suffers less and less diminution of intensity by each additional plate; so that when a certain number has been attained, the intensity received by the eye, or on a screen, is practically unaffected by increasing their number.

Fig. 8 is a general representation of such a pile of plates viewed edge-ways. The plates are secured in a brass frame, and the whole supported on a stand.

The experiment described above, in which the rays reflected from the pile of glass plates are extinguished by the analyser when in one position, while those which have been transmitted are extinguished when the analyser is in a position at right angles to the former, shows that the vibrations of the reflected and refracted rays, so far as they become polarised, are at right angles to one another. And further, if these rays be severally examined with a plate of tourmalin, it will be found that the vibrations of the reflected ray are executed in a direction perpendi-

[1] Plates of the thinnest description are the best. Two or three give good effects; but if the surfaces lie parallel and the glass be highly transparent, the number may be advantageously increased to 10 or even 12.

cular to the plane of incidence, and those of the refracted ray in a direction parallel to that plane.

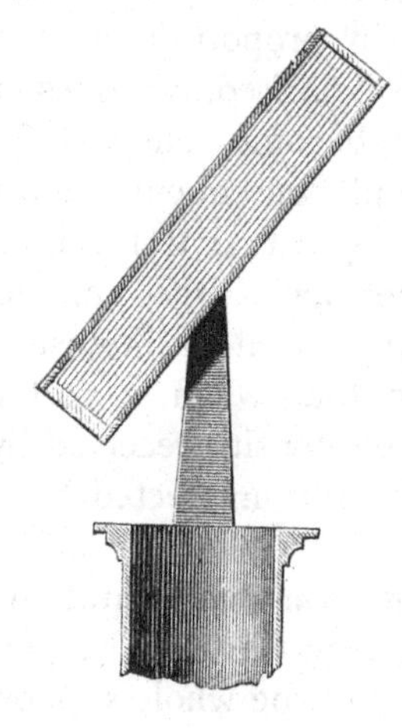

FIG. 8.

The same general reasoning as that used in the case of tourmalin plates will serve, if not as actual proof, at all events as illustration in this case. Thus, suppose that a ray whose vibrations are perpendicular to the plane of incidence—that is, parallel to the reflecting surface—fall upon a plate of glass, then there is no apparent reason why a change in the angle of incidence should modify the reflection and refraction, so far as they depend directly upon the direction of the vibrations. The vibrations cannot undergo any change of direction on one side rather than on the other by incidence on a surface to which they are parallel, and will consequently remain parallel to themselves even when the incidence has taken place. And since the reflected and refracted rays both lie in the plane of incidence, the vibrations (which are perpendicular to that plane, and consequently to every line in it) will fulfil the optical condition of being perpendicular to the rays in question. But if the vibrations of the incident ray take place in the plane of incidence, it is difficult to conceive that the results of reflexion and refraction should be unaffected by a change in the angle of incidence. There are two mathematical and mechanical principles which, when applied to the case of vibrations in the plane of incidence, lead to the conclusion that, if the ray be incident at such an angle that the reflected and refracted rays are perpendicular to one another, there can be no reflected ray.

A general explanation of this very curious result seems difficult; but the following considerations may perhaps tend to elucidate the subject. Reflexion is generally, perhaps always, accompanied by refraction. Bodies are visible in virtue of rays which, after reflexion from their surface, meet the eye. But the natural colours of bodies so seen are due to rays which are not reflected until they have penetrated to some, although inconsiderable, depth below the actual surface. During this penetration the light has been deprived of certain of its component rays, and emerges as a reflected beam coloured with the remaining or complimentary tint. Light reflected merely from the outer surface of bodies is in general colourless. A homogeneous coloured liquid in a perfectly black vessel would show nothing but colourless reflexion from the surface ; but mix (say) chalk, so as to make it muddy, and colour will be seen. The colourless reflexion from polished surfaces is perhaps only a limiting case in which the penetration is a minimum. If this be so, we may fairly conclude that refraction is the ruling feature of the phenomenon, and that it in some sense precedes reflexion. With the change of direction of the ray involved in refraction it is in the highest degree probable that a change of direction of the vibrations (supposed always to be in the plane of incidence) will be also involved. The simplest supposition would be that the vibrations within the medium are perpendicular to the refracted ray; and that the intensity of the reflected light is due to that part of them which can be resolved in a direction perpendicular to that of the reflected ray. If, therefore, the refracted and the reflected rays be perpendicular, so also will be their vibrations, and consequently no part of the vibrations constituting the former can be resolved in the direction requisite for the latter. In other words, there will be no reflected ray.

The above remarks give, it must be admitted, no mechanical theory of reflexions, nor indeed do they pretend to be even a rough explanation of the facts. They merely amount to this: if reflection depends primarily upon refraction, and the known law of reflexion obtains independently of all questions of polarisation, then, when the incident vibrations take place in the plane of incidence, no reflected ray whose direction is perpendicular to that of the refracted ray can be produced.

CHAPTER II.

DOUBLE REFRACTION—POLARISCOPES.

WE next come to the subject of polarisation by double refraction. There are a large number of crystals which have the property of generally dividing every ray which passes through them into two. But the extent of separation of the two rays varies with the direction of the incident ray in reference to the natural figure of the crystal. In every double refracting crystal there is at least one, and in many there are two, directions in which no such separation takes place. These directions are called optic axes. The relations between the forms of crystals and their optic axes, and optical properties arising therefrom, will be explained later.

Of such crystals Iceland spar is the most notable instance. If we take a block of such spar split into its natural shape, a rhombohedron, Fig. 9, and for convenience cut off the blunt angles by planes perpendicular to the line joining them *a b*, it will be seen that a ray of light transmitted perpendicularly to these planes, that is, parallel to the line joining the blunt angles, is not divided. In fact, the image either of the aperture of the lantern projected on a screen, or of an object seen by the eye in the direction in question, appears single, as if passed through a block of glass. The direction in question (viz. the line *a b* itself, and

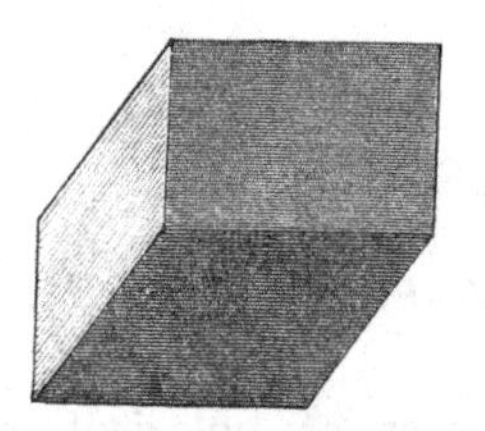

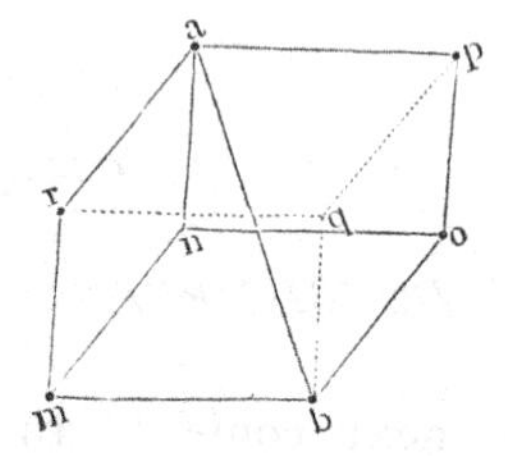

FIG. 9.

all lines passing through any part of the crystal parallel to *a b*) is called the optic axis of the crystal. If, however, the crystal be tilted out of this position in any direction, it will be seen, by the appearance of two images instead of one, that the rays are divided into two. The angular divergence of the two sets of rays, or, what comes to the same thing, the separation of the two images, depends upon the angle through which the crystal has been turned ; or, as it may also be expressed, upon the angle between the directions of the incident ray and the optic axis of the crystal. When this angle amounts to a right angle, the separation is at its greatest ; and if the crystal be still further turned, the images begin to come together again until, when it has turned through another right angle, they coincide.

This process of separation, or doubling the rays, is called double refraction. And the following experiment will show that one set of rays follows the ordinary law of refraction, while the other follows a different law. The image produced by the first set of rays is, in consequence, called the ordinary, and that produced by the second the extraordinary image. Let us now take a sphere of Iceland spar, which will act upon the rays issuing from the lamp as a powerful lens. In every position in which it is placed it produces two images on the screen; but in one position the two images are concentric, and differ only in this, that one is larger than the other. The direction in which the light is then passing is that of the optic axis; and it is to be observed that, although there is a difference in the magnifying of the two images, there is still no divergence of rays, or separation of images in the sense used before. In fact, if we suppose the curvature of the lens to be gradually diminished, we should find the difference of the sizes of the two images, as well as the absolute size of both, diminish; until when the surfaces of the lens became flat, the difference would vanish, and the two images would absolutely coincide.

This difference in the size of the images shows, moreover, a very important property of double refracting crystals. The amount of refraction produced by a transparent medium standing in air depends, as is well known, upon the velocity with which a ray of light traverses the medium compared with that with which it traverses air. The smaller the velocity in the medium, the greater the refraction. The greater the refraction, the greater the magnifying power of a lens constructed of that medium. Hence in the two concentric images we can at once point to the system of rays which has traversed the crystal at a lower velocity than the other.

Let us now turn the crystal round into some other position, so that the direction of the optic axis shall no longer coincide with that of the rays from the lamp or from the object. During this process one of the images, the larger, remains stationary, as would be the case with the single image, if we had used a sphere of glass. This, therefore, is the ordinary image. The other shifts about, separating itself from the first, until the crystal has been turned through half a right angle, and then drawing back again until the crystal has swept round through a complete right angle. This is, consequently, the extraordinary image.

It will be noticed that when the sphere has been turned through a right angle, the extraordinary image is no longer circular, but elliptical, and that the major axis of the ellipse lies in the direction in which the motion has taken place, that is, perpendicular to the axis about which the sphere has been turned. This is due to the fact, shown above, that the nearer the direction of the incident rays to that of the optic axis, the less the divergence between the ordinary and the extraordinary rays. The distortion of the image when the sphere has turned through half a right angle is due to the difference of angles between the optic axis and the rays which enter the crystal on one side and on the other of the central ray of the beam coming from the lamp.

That the rays forming each of the images are polarised, and that the direction of their polarisation is different, is easily shown by interposing a plate of tourmalin or other polarising instrument between the lamp and the sphere of spar. But inasmuch as the polarisation in many positions of the sphere is far from uniform, the phenomenon becomes rather complicated; and the character of the polarisation of the two images is better studied by using flat instead of curved surfaces for separating the rays.

For the purpose in question there is, perhaps, no better instrument than the double-image prism. This consists of a combination of two prisms, one of Iceland spar, so cut that the optic axis is parallel to the refracting edge; the other of glass, and usually having a refracting angle equal to that of the spar. The rays passing through the crystal prism being perpendicular to the optic axis, undergo the greatest separation possible. And the chromatic dispersion caused by that prism is corrected or neutralised entirely in the case of the extraordinary, and nearly so in that of the ordinary ray, by the glass prism which is placed in a reversed position. In this arrangement the extraordinary image occupies the centre of the field, and remains fixed when the double-image prism is made to revolve in a plane perpendicular to the incident rays; while the ordinary image is diverted to a distance from the centre, and revolves in a circle about that centre, when the prism revolves.

If the nature of the light in the two images thus formed be examined by any polarising instrument, it will be found that it is polarised in both cases; and that the vibrations in the one image are always perpendicular to those in the other. And in particular the vibrations in the extraordinary image are parallel, and those in the ordinary are perpendicular to the optic axis.

On these principles polarising and analysing instruments have been constructed by various combinations of wedges or prisms of Iceland spar, the details of which it is not necessary to describe in full. But the general problem, and object proposed, in all of them has been to cause such a separation of ordinary and extraordinary rays, that one set of rays may, by reflexion or other methods, be further diverted and afterwards thrown altogether out of the field of view. This done, we have a single beam of completely polarised light and a single image produced from it.

One such instrument, however, the Nicol's prism, on account of its great utility and its very extensive

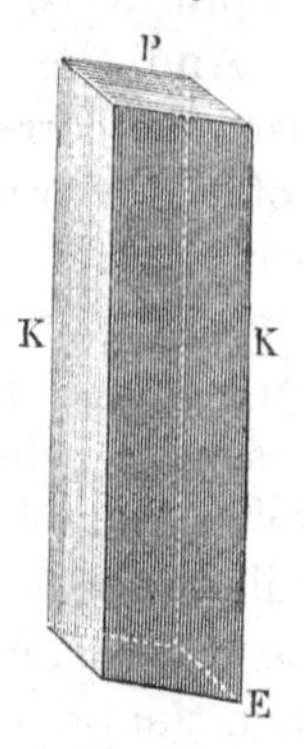

FIG. 10.

use, deserves description. A rhombohedron of Iceland spar double of its natural length is taken (see Fig. 10); and one of its terminal faces P, which naturally makes an angle of 71° with the blunt edges K, is cut off obliquely so as to give the new face, say P′ (not given in the figure), an inclination of 68° to the edges K. The whole block is then divided into two by a cut through the angle E in a direction at right angles to the new face P′; the faces of this cut are then care-

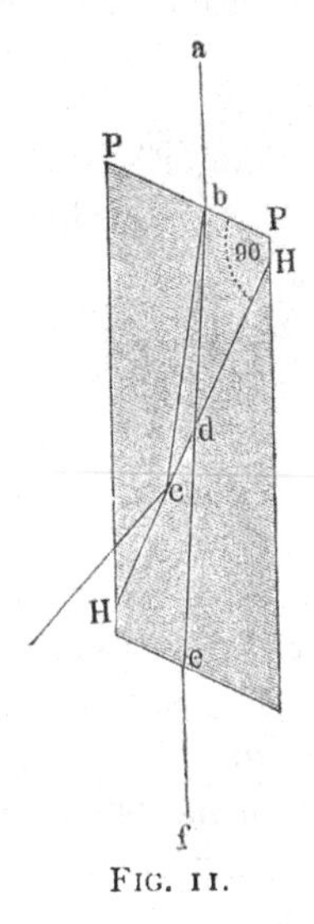

FIG. 11.

fully polished, and cemented together again in their original position with Canada balsam. Fig. 11 represents a section of such a prism made by a plane passing through the edges K (Fig. 10). A ray entering as *a b* is divided into two, viz. *b c* the ordinary, and *b d* the extraordinary. But the refractive index of the Canada balsam is 1·54, *i.e.* intermediate between that of the spar for the ordinary (1·65) and the extraordinary (the minimum value of which is 1·48) rays respectively; and in virtue of this the ordinary ray undergoes total reflexion at the surface of the balsam, while the extraordinary passes through and emerges ultimately parallel to the incident ray. Fig. 12 shows an end view of a Nicol's prism, the shorter diagonal in the direction of vibration of the emergent polarised ray.

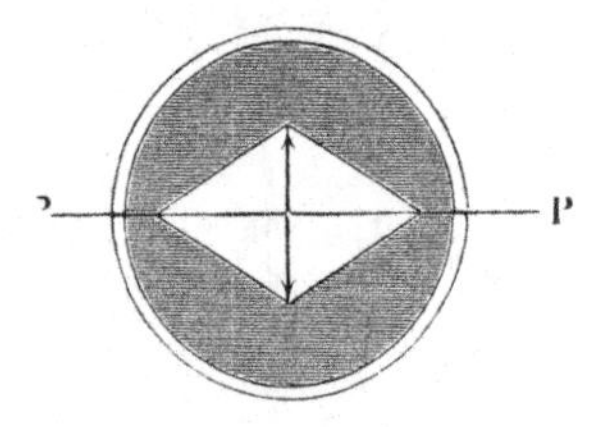

FIG. 12.

Two such instruments, when used together, are respectively called the "polariser" and the "analyser," on account of the purposes to which they are put. These, when placed in the path of a beam of light, give rise to the following phenomena, which are, in fact, merely a reproduction in a simplified form of what has gone before.

When polariser and analyser are placed in front of one another with their shorter diagonals parallel, that is, when the vibrations in the image transmitted by the one are parallel to those in the image transmitted by the other, the light will be projected on the screen exactly as if only one instrument existed. If, however, one instrument, say the analyser, be turned round, the light will be seen to fade in the same way as in the case of the tourmalin plates; until, when it has been turned through a right angle, or, as it is usually expressed, when the polariser and analyser are crossed, the light is totally extinguished.

In the complete apparatus or polariscope, we may incorporate any system of lenses, so that we may make use of either parallel or convergent light, and finally focus the image produced upon the screen or upon the retina. At present we shall speak only of the phenomena of colour produced by crystal plates in a parallel beam of polarised light—chromatic polarisation, as it is called, with parallel light.

Various forms of polariscopes have been devised, whereof the three described below may be regarded as the most important.

Fig. 13 is an elevation of one of them. When used in its simplest form, the frame F carries a plate of black glass, which is capable of revolving about pivots in the uprights. The positions of the source of light and of the frame must be adjusted so that the plate will receive the incident light at the polarising angle, and reflect it in the direction of the eye-piece, which contains a Nicol or other analyser. The objects to be examined are to be placed on the diaphragm E.

This instrument may be converted into another form, due to Norremberg, by placing a silvered mirror horizontally at H. The plate of black glass must be removed from the frame F, and a plate of transparent glass substituted for it, which must be so inclined that the light falling upon it shall be reflected at the polarising angle perpendicularly towards the horizontal mirror. The object may be placed on the diaphragm E as before. But it may also be placed on the diaphragm D below the polarising plate F, and in that case the eye will receive the polarised ray reflected

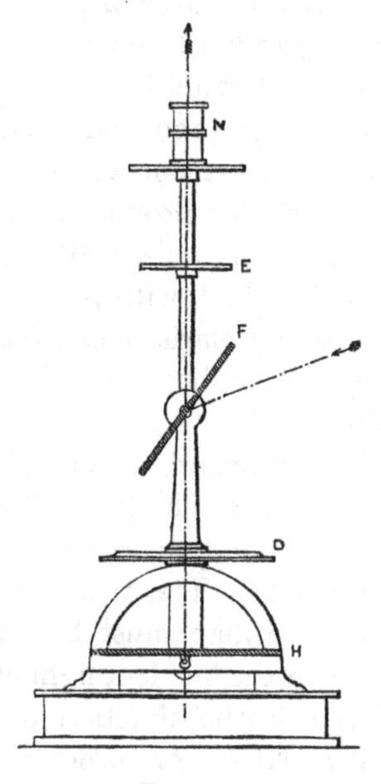

FIG. 13.

from the mirror; and the polarised ray will have passed, before it reaches the eye, twice through the crystalline plate placed between the mirror and the polariser. The result is the same as if, in the ordinary apparatus, the polarised ray had passed through a plate of double the actual thickness. If the plate does not fill the entire field of view, two images of the plate will be seen, the one larger, as viewed directly, the other smaller, as viewed after reflexion from the horizontal mirror; the first will show the tint due to the actual thickness of the crystal, the other that due to a plate of the same crystal, but of double the thickness.

CHAPTER III

CHROMATIC POLARISATION—THE WAVE THEORY.

WE now proceed to the consideration of the colours produced by plates of crystal when submitted to the action of polarised light. A crystal very commonly used for this purpose is selenite or hydrated sulphate of lime, which is readily split and ground into flat plates of almost any required thickness. If such a plate be placed between the polariser and analyser when crossed, it will be found that there are two positions, at right angles to each other, in which, if the selenite be placed, the field will remain dark as before. The selenite is, in fact, a doubly refracting crystal, and although neither of the rays to which it gives rise follows the ordinary law, yet they may, for sake of distinction, be respectively called the ordinary and extraordinary. The positions in question are those in which the plane of vibration of the ordinary ray coincides with that of the polariser (or analyser), and that of the extraordinary ray with that of the analyser (or polariser). In every other position of the selenite, and notably when it has turned through 45° from either of the positions before mentioned, or neutral positions, as they may be called, light passes through, and the field becomes bright. If the thickness of the selenite be considerable, the field when bright will be colourless; but if it be inconsiderable, say not more than that of a thin visiting card, the field will be brilliantly coloured with tints depending upon the thickness of the plate.

Supposing however that, the selenite remaining fixed, the analyser be turned round, we shall find that, in the first place, the colour gradually fades as before; until, when the analyser has been turned through 45°, all trace of colour is lost. After this, colour again begins to appear; not, however, the original tint, but its complementary. And, in fact, there is no more sure way of producing colours complementary to one another than that here used. A general explanation of this change of colour is already furnished by our former experiments. Doubly refracting crystals generally, in the same way as Iceland spar, divide every ray, and consequently every beam of light, which passes through them, into two, so that of every object seen through them, or projected through it on to a screen, two images are produced. These two, being parts of one and the same beam of light, would, if recombined, reproduce the original beam; and the same is, of course, the case with the two images. This may be rendered visible by using the double-image prism as an analyser, and throwing both images on the screen together. It will then be seen that whatever changes the colours undergo when the prism is turned round, the overlapping part of the images is white; and their tints are therefore complementary.

If the distance of the prism be so adjusted that the images overlap, it will be found that, when both are visible, the part which they overlap is always white, whatever be the thickness of the plate used. And an instructive experiment may be made by interposing an opaque object in the path of the beam of light, so that its shadow may fall upon the part of the field common to the two images. The shadow will of course intercept the light forming each of the images, and will consequently appear double. Suppose that the two images are coloured red and green respectively; then one of the shadows will be due to the shutting off of the red light, and the other to that of the green. But in the first case the space occupied by the shadow will be still illuminated by the green light, and in the second by the red. In other words, neither of the two shadows will be black; one will be green and the other red. If in any part of their extent the two shadows overlap, the part common to the two, being deprived of both red and green light, will be black.

But in order to explain how it comes to pass that colour is produced at all, as well as to find a more strict proof that the colours of the two images are complementary, we must have recourse to some considerations based upon the Wave Theory of Light. And, first, as to the mode in which waves may be produced.

Consider a row of balls lying originally in a horizontal straight line. Let the balls start one after another and vibrate at a uniform rate up and down. At each moment some will be at a higher, others at a lower level, at regular intervals in a wave-like arrangement; the higher forming the crests, the lower the hollows of the waves. The distance from crest to crest, or from hollow to hollow, is called the *wave length.* The distance from crest to hollow will consequently be half a wave length. This length will be uniform so long as the vibrations are executed at a uniform rate.

Each ball in turn will reach its highest point and form a crest; so that the crests will appear to advance from each ball to the next. In other words, the waves will advance horizontally, while the balls vibrate vertically.

If the row of balls were originally arranged in a wave form of the same amplitude, and caused to vibrate in the same way as before, those on the crests would vibrate wholly above, and those in the hollows wholly below the middle line. When the balls originally on the crests rise to their highest positions, those in the hollows will fall to their lowest positions, and the height of the wave will consequently be doubled. When the balls originally at the crests fall, those in the hollows will rise, both to the middle line; and the wave will consequently be annihilated. The first of these corresponds to a condition of things wherein the crests of the new wave motion coincide with those of the old, and the hollows with the hollows; the second to that wherein the crests of the new coincide with the hollows of the old, and *vice versâ.*

Hence, when two sets of waves are coincident, the height of the wave or extent of vibration is doubled; when one set is in advance of the other by half a wave length, the motion is annihilated. The latter phenomenon is called *interference.* When one set of waves is in advance of the other by any other fraction of a wave length, the height of the wave, or extent of vibration, is affected, but not wholly destroyed; in other words, partial interference takes place. The distance whereby one set of waves is in advance of another is called the *difference of phase.*

The Wave Theory of Light consists in explaining optical phenomena by vibrations and waves of the kind above described. And according to that theory the direction in which the waves move is the direction of propagation of the ray of light.

The intensity of light depends upon the extent of the vibrations or the height of the waves; the colour upon the number of vibrations executed in a given interval of time. And since throughout any uniform medium the connexion of the parts and the rate of propagati n may be considered to be uniform, it follows that the waves due to the slower vibrations must be longer than those due to the more rapid. Hence the colour of the light may be regarded as depending upon the wave length.

The substance to the vibrations of which light is supposed to be due is an elastic fluid or medium pervading all space, and even permeating the interior of all bodies. A full statement of the reasons which have led philosophers to make this hypothesis would involve considerations derived from other sciences beside optics, and would be out of place here. But it may still be pointed out that one strong argument is furnished by the fact of the transmission of light from the sun and from the fixed stars through space, where no atmosphere or gases as known here can be conceived to exist. That the light so traversing interstellar space must be transmitted by a material substance is a fundamental proposition of mechanical philosophy; and the hypothesis of the ether simply consists in attributing to the substance or medium the property of elasticity (a property possessed in a greater or less degree by all known bodies), without assuming anything else whatever as to its nature or relation to other substances.

In the illustrations of wave motions given above, the balls would represent successive portions or molecules of the ether; and the means whereby the motion of one molecule is transmitted to its neighbour is their mutual attraction, or the elastic cohesion attributed to the whole medium in the hypothesis in question.

The difference between ordinary and polarised light has been explained above; and the mechanical contrivances devised for representing wave motion always have reference only to polarised light. But as this is the subject with which we are here concerned, the limitation in question is not of consequence. A variety of instruments have been constructed for showing the motions of the particles of ether in the plane, circular, and elliptical vibrations of light; and Sir Charles Wheatstone has invented a machine which represents to the eye in what manner plane waves in the same or in rectangular planes combine, so as to form every variety of plane, circular, or elliptical waves. It also exhibits the resultants of two series of circular or elliptical waves, according to their differences of phase; and it further shows that when two circular waves, one right-handed and the other left-handed, are combined, a series of plane waves is formed, the azimuth of which changes with the difference of phase.

In plane-polarised light, such as is produced by tourmalin plates, by double refraction in Iceland spar, &c., the vibrations are rectilinear, and are executed in one and the same plane throughout the entire length of the ray. In circularly polarised light the vibrations are all circular, and throughout the same ray the motion is performed in the same direction. In elliptically polarised light the vibrations are all elliptical, the ellipses are all similarly placed, and the motion is in the same direction for the entire ray. These are the only known forms of polarisation, and indeed they are the only forms compatible with the usual, simplest assumption respecting the elasticity of the ether.

These general considerations being premised, we are in a position to trace the course and condition of a ray of light issuing from the lamp or other source, and traversing first the polarising Nicol's prism; secondly, the plate of doubly refracting crystal; thirdly, the analysing Nicol.

The vibrations of the ray on leaving the polariser are all restricted to a single plane. On entering the plate of doubly refracting crystal, every ray is divided into two, whose vibrations take place in planes perpendicular to one another. The angular position of these planes about the axis of the beam of light is dependent upon the angular position of the crystal plate about its centre. The two sets of rays traverse the crystal with different velocities, and therefore emerge with a difference of phase. The amount of this difference is proportional to the thickness of the plate. On entering the analyser the vibrations of each pair of rays are resolved into one plane; and are then in a condition to exhibit the phenomena of interference. If the plane of vibration of the analyser be parallel to one of those of the plate, that ray will be transmitted without change; the other will be suppressed. In any other position of the analyser those monochromatic rays (spectral components of white light) whose difference of phase most nearly approaches to half a wave length will be most nearly suppressed; and those in which it approaches most nearly to a whole wave length will be most completely transmitted. The amount of light suppressed increases very rapidly in the neighbourhood of the ray whose difference of phase is exactly a half-wave length; so that with plates of moderate thickness a single colour only may, speaking roughly, be considered to be suppressed. This being so, the beam emergent from the analyser will be deprived of that colour, and will in fact consist of an assemblage of all others; or, in other words, will be of a tint complementary to that which has been extinguished.

Next, as regards the colours of the two images—that is, the two which are formed either simultaneously by a double-image prism, or successively by a Nicol in two positions at right angles to one another. In the first place, it is to be remembered that the two sets of vibrations into which the selenite has divided the polarised ray are at right angles to one another;

secondly, that one set is retarded behind the other through a certain absolute distance, which is the same for every ray, and consequently through a distance which is a different fraction of the wave length for each colour; thirdly, that these two are re-combined or "resolved" in a singular direction in each image by the analyser.

This being so, bend two wires in the following form:—

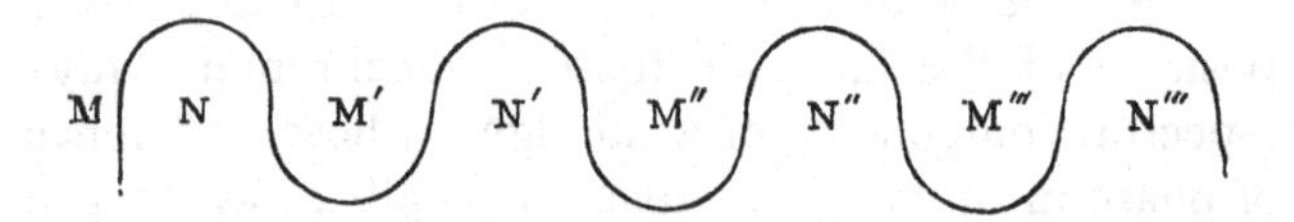

and place them at right angles to one another about their middle line M N M′ N′ . ., so that the points M of the two wires coincide, and likewise N, and so on. This will represent the condition of the vibrations as they emerge from the selenite, when the plate is of such a thickness as to cause a retardation equal to one or to any whole number of wave lengths. Turn the wires about their middle lines M N M′ N′ until they meet half-way, *i.e.* in a position inclined at 45° to their original directions; this will represent the vibrations as resolved by the analyser in one image. Turn the wires about their middle line as before, but in reversed directions, until they meet in a position at right angles to the former; this will represent the vibrations as resolved by the analyser in the other image. On looking at the wires when so brought together, it will be found that in one case the crests fall upon the crests and the hollows upon the hollows, so that the vibrations combine to increase the intensity of the light. In the other case the crests fall upon the hollows and the hollows upon the crests, so that the vibrations interfere and completely neutralise one another.

The same principle would obtain if we shifted one wire along the middle line so that the points M of the two wires no longer exactly coincide. This would represent the condition of the vibrations as they emerge from the selenite when the plate is of such a thickness as to cause a retardation of a fraction of a wave length equal to the amount of shift. And on turning the wires as before, we should find that in one image the waves partially combine, and that in the other they partially interfere. The shifting of the wires would represent either the effect of plates of different thickness upon waves of the same length, *i.e.* rays of the same colour ; or that of a single plate on waves of different lengths, *i.e.* on rays of different colours. From these considerations we may conclude that the rays which are brightest in one image are least bright in the other ; or, in other words, that the colours of the two images are complementary.

It has been remarked that the colour produced by a plate of selenite depends upon the thickness of the plate. In fact, the retardation increases with the thickness, and consequently, if, for a given thickness, it amounts to a half-wave length of the shortest (say violet) waves, for a greater thickness it will amount to a half of a longer (say green) wave, and so on. And if, instead of a series of plates of different thicknesses, we use a wedge-shaped plate, the entire series of phenomena due to gradually increasing retardation will be produced. This is easily seen to consist of a series of tints due to the successive extinction of each of the rays, commencing with the violet and ending with the red. And the tints will consequently have for prevailing hues the colours of the spectrum in the reverse order. This series of colours will be followed by an almost colourless interval, for which the retardation is intermediate between a half-red wave length and three half-violet wave lengths. Beyond this again the series of colours will recur; and the same succession is repeated as the wedge increases in thickness. It will, however, be observed that the colours appear fainter each time that they recur, so that when the thickness reaches a certain amount (dependent upon the nature and retarding power of the crystal) all trace of colour is lost.

It is not difficult to account for this gradual diminution in the intensity of the colours if, by means of a diagram, we examine the mode in which the waves of various lengths interfere with one another; but spectrum analysis furnishes an explanation which is perhaps more easy of general apprehension. If the light emerging from the analyser be examined by a spectroscope, it will be found, in the case of a plate giving the most vivid colour, that the spectrum presents a dark band indicating the colour which has been extinguished. On using thicker and thicker plates, the band will be found to occupy positions nearer and nearer to the red end of the spectrum, until the band finally disappears in the darkness beyond the least refrangible rays that are visible to the eye. If the analyser be turned round, the band will gradually lose its characteristic darkness, until, when the angle of rotation has reached 45°, the band will have disappeared altogether. The spectrum is then continuous, and when recompounded will give white light. This corresponds to the fact noticed before, that when the analyser is turned round, the colour given by a selenite plate fades, and finally disappears when the angle of rotation amounts to 45°. If the rotation be continued, a band reappears, not, however, in its original position, but in the part of the spectrum complementary to the former.

If the thickness of the plate be further increased, two bands will be seen instead of one ; with a still greater thickness there will be three bands, and so on indefinitely. The total light, then, of which the spectrum is deprived by the thicker plates is taken from a greater number of its parts ; or, in other words, the light which still remains is distributed more and more evenly over the spectrum, and consequently at each recurrence of the tints the sum total of it approaches more and more nearly to white light.

The following experiment will be found very instructive. Take two wedges of selenite or other crystal, and having crossed the polariser and analyser, place the two wedges side by side in the field of view so as to compare the tints produced by the two. Then place one over the other, first with the thick end of the one over that of the other, next with the thick end of the one over the thin end of the other. If the two plates are exactly similar, the combination in the first instance will be equivalent to a single wedge whose refracting angle is double that of a single wedge; and the number of bands produced will consequently be doubled. In the second combination the angles of the wedges will compensate one another, and the result will be equivalent to a uniform plate whose thickness is equal to the sum of the mean thicknesses of the wedges. The field will then be coloured with a uniform tint, viz. that due to a plate of the thickness in question.

By making use of the principle that the colour produced depends upon the thickness of the plate, selenites have been cut of suitable shapes and thicknesses, so as to produce coloured images of stars, flowers, butterflies, and other objects.

The phenomena exhibited by selenite are also produced by other crystals, but the facility with which plates of the former substance can be obtained causes them to be generally used in preference to others. There is, however, a peculiar class of crystals, of which quartz, or rock crystal, is the most notable, which gives rise to effects different from those hitherto described.[1]

[1] See Proceedings of the Royal Institution, 1873.

CHAPTER IV.

CIRCULAR POLARISATION.

IF a ray of light pass through a plate of quartz which has been cut perpendicularly to the axis, or line parallel to the length of the crystal (hexagonal prism), it is as usual divided into two; but the vibrations in each ray, instead of being rectilinear and at right angles to one another, are circular and in opposite directions. That is to say, if the motion of vibration in one ray is directed like the hands of a clock, that in the other is directed in the opposite sense; and the light in each ray is then said to be circularly polarised. The motion of a series of particles of ether, which when at rest lie in a straight line, is circular, and, as in plane polarisation, successive; and consequently at any instant during the motion such a series of particles will be arranged in a helix or corkscrew curve. The sweep of the helix will follow the same direction as that of the circular motion; and, on that account, a circularly polarised ray is spoken of as right-handed or left-handed, according to the direction of motion. A right-handed ray is one in which, to a person looking in the direction in which the light is moving, the plane of vibration appears turned in the same sense as the hands of a watch. Or, what is the same thing, to a person meeting the ray, it appears turned in the opposite sense, viz. that in which angles when measured geometrically are usually reckoned as positive.

The question, however, which mainly concerns us is the condition of the vibrations after emerging from the plate of quartz and before entering the analyser. In the passage of the ray through the plate the ether is subjected to a double circular motion, one right-handed, the other left-handed ; but, as one of these motions is transmitted with greater velocity than the other, it follows that at any given point and at the same instant of time one of the revolutions will, in general, be more nearly completed than the other, or, to use an expression adopted in plane polarisation, there will be a difference of phase. The motions may be represented by two clock hands moving at the same rate in opposite directions, and the difference of phase by the angle between them when one of them is in the position from which angles are reckoned. As both are supposed to move at the same rate, they will have met in a position midway between their actual positions ; and if we consider a particle of the ether (say) at the extremity of the clock-hands, it will be solicited when the hands are coincident by forces producing two opposite circular motions. Now, whatever may have been the forces or structural character within the crystal whereby this double circular motion is perpetuated, it is clear that when the ray emerges into air the particle of ether immediately contiguous to the surface of the crystal will be acted on by two sets of forces, one whereby it would be caused to follow the right-handed and the other the left-handed rotation. Each of these may, as is well known, be represented by a pair of forces, one directed towards the centre of the circle, the other in the direction of the motion and at right angles to the first, or, to use geometrical language, one along the radius and towards the centre, the other along the tangent and in the direction of the motion. The two forces acting along the tangent, being in opposite directions, will neutralise one another, and the resultant of the whole will, therefore, be a force in the direction of the centre. The particle in question, and consequently all those which, following in succession, serve to compose the entire ray until it enters the analyser, will vibrate in the direction of the diameter drawn through the point under consideration; or, to express it otherwise, the ray will be plane-polarised, and the plane of vibration will be inclined to the plane from which angles are measured by an angle equal to half the difference of phase on emergence due to the thickness of the crystal. The retardation being very nearly the same, absolute quantity for all rays will, as in the case of plane polarisation, be a different fraction of the wave length for rays of different colours, and will be greater for the shorter waves than for the longer. Hence the planes of vibration of the different coloured rays, after emerging from the quartz, will be differently inclined. Each ray will therefore enter the analyser in a condition of plane polarisation; and if the analyser be turned round, it will cross the vibrations of the various coloured rays in succession, and extinguish each of them in turn. Each of the images will consequently exhibit a gradual change of colour while the analyser is being turned; and the tints will be, as explained before, complementary to those which are successively extinguished. For a given plate of quartz the order of the tints will be reversed when the direction of rotation of the analyser is reversed. But it should be here explained that there are two kinds of quartz, one called right-handed and the other left; and that, for a given direction of rotation of the analyser, these cause the colours to follow one another in opposite orders. A similar effect is produced by turning the polariser round in the opposite direction.

The angle of rotation of the plane of vibration for any particular colour varies, as stated above, with the thickness of the plate; while for a given thickness it increases nearly as the square (product of the quantity into itself) of the wave length decreases. In mathematical language, it varies approximately inversely as the square of the wave length. If this law were accurately true, the product of the angles of rotation into the square of the corresponding wave lengths (λ) would be the same for all rays. The following are some measurements made by Brock, with a quartz plate one millimetre thick, which show that the law may be considered as true for a first approximation.

Rays.	Rotations.	Rotations $\times \lambda^2$.
B	15° 18′	7,238
C	17° 15′	7,429

Rays.	Rotations.	Rotations × λ^2.
D	21° 40′	7,511
E	27° 28′	7,596
F	32° 30′	7,622
G	42° 12′	7,842

If the colours exhibited by a plate of quartz when submitted to polarised light be examined by a spectroscope, in the way described when we were speaking of selenite, the spectrum will be found to be traversed by one or more dark bands, whose position and number depend upon the thickness of the plate. But there will be this difference between plane and circular polarised light, that if the analyser be turned round the bands will never disappear, but will be seen to move along the spectrum in one direction or the other, according as the plate of quartz be right-handed or left-handed, and according to the direction in which the analyser is turned. This is, in fact, identical with the statement made before, that the analyser in its different positions successively crosses the plane of vibration of each ray in turn, and extinguishes it.

This being so, it is clear that a change of colour exhibited by a quartz plate, when submitted to plane-polarised light and examined with an analyser, forms a test of a change in the plane of original polarisation. And if the plate be composed of two parts, one of right-handed the other of left-handed quartz, placed side by side, any change in the plane of polarisation will affect the two parts in opposite ways. In one part the colours will change from red to violet; in the other from violet to red. At two positions of the polariser, or analyser, the colours must be identical.

With plates, as usually cut, one of these identities will be in the yellow, the other at the abrupt passage from violet to red, or *vice versâ*. In the latter case the field appears of a neutral tint—*teinte sensible* or *teinte de passage*, as the French call it—and the slightest change in the plane of polarisation exhibits a marked distinction of colour, one part verging rapidly to red, the other to violet. This arrangement is called a biquartz, and affords a very delicate test for determining the position, or change of position, of the plane of polarisation, especially in cases where feebleness of light or other circumstances interfere with the employment of prismatic analysis.

If the thickness of the plate be such that the difference of rotation of the planes of vibration of the rays corresponding to the two ends of the visible spectrum (or, as it is sometimes termed, the "arc of dispersion") be less than 180°, there will be one dark band in the spectrum; because there can then be only one plane of vibration at a time at right angles to that of the analyser. If the arc of dispersion is greater than 180° and less than 360°, there will be two bands. And so on for every 180° of dispersion.

This mode of examination by means of prismatic analysis is the most accurate yet devised for measuring the angle of rotation produced by circular polarisation; especially if solar light be employed, and the fixed lines used to form a scale of measurement.

The property of circular polarisation is, however, not confined to quartz.

The following crystals are known as having the property of rotatory polarisation.

In the cubic system—

Sodium chlorate.
" bromate.
" iodate.

In the tetragonal system—

Strychnine sulphate crystallised with 13 equivalents of water.

Æthylodiamine sulphate.

In the hexagonal system—

Quartz.

Cinnabar.

Sodium periodate ($NaIO_4 + 2H_2O$).

Lead hyposulphate ($PbS_2O_6 + 4H_2O$).

Benzil.

(Possibly also some other hyposulphates.)

Sir John Herschel was the first to observe a remarkable connexion between the external form and the direction of rotation produced by different specimens of quartz. The normal form of the crystal is that of a hexagonal prism, terminated at each end by hexahedral pyramids. The angles at the ends of the edges of the prism are sometimes cut off by little facets, which in some cases are inclined to the right, in others to the left, when the prism is held vertically. But when the upper facet is inclined to the right, the lower facet is inclined to the left, and *vice versâ*; and the direction of the rotation produced by the crystal follows the inclination of the upper facet.

In the language of crystallography these facets are termed hemihedral planes. And from the fact that two planes on the same edge of the prism are always inclined in opposite directions, it follows that a right-handed and a left-handed specimen cannot be geometrically superposed ; one is in fact the reflected image of the other. The formal relation between the two kinds has on this account been termed hemihedral hemi-symmetry.

Quartz and sodium periodate and lead hyposulphate are the only crystals in which right and left-handed rotatory polarisation has been identified with hemihedral hemi-symmetry.

The following list is given by Verdet. The angles have reference to the red rays given by a plate of glass coloured with oxide of copper, and are affected with the sign + in the case of right-handed, and with − in the case of left-handed rotation. The length of the column of the solution is in every case one decimetre.

Essence of	turpentine	−29°·6
"	lemon	+55°·3
"	bergamot	+19°·08
"	bigarade	+78°·94
"	aniseed	− 0°·70
"	fennel	+13°·16
"	carroway	+65°·79
"	lavender	+ 2°·02
"	peppermint	−16°·14
"	rosemary	+ 2°·29
"	marjorum	+11°·84
"	sassafras	+ 3°·29
Solution of	sugar 50 per cent.	+33°·64
"	quinine 6 per cent. in alcohol	−30°

It will be noticed that the rotatory power of all these substances is much less than that of quartz.

A mixture of liquids, one or both of which is active, generally exhibits a rotatory action represented by the sum or difference of their separate powers (a neutral liquid being considered to have a power represented by 0); but this law is true only when no chemical action takes place between the elements of the mixture. Saccharine solutions vary not only in the amount but also in the character of their power of rotation; thus cane sugar is right-handed, but grape sugar left-handed.

The property in question has been turned to practical use by employing the rotatory power of a saccharine solution as a measure of the strength of the solution. For this purpose a tube containing the solution to be examined is placed between two Nicol's prisms. The simple fact of circular polarisation is proved by a feeble exhibition of the phenomena shown by a plate of quartz cut perpendicularly to the axis. But for accurate measurement various expedients have been adopted. If a biquartz be inserted behind the analyser (the end of the apparatus next the eye being considered the front), then for a certain position of the analyser the two halves will appear of the same colour. When the tube for examination is inserted, the similarity of colour will be disturbed; and the angle through which, right or left, the analyser must be turned in order to restore it will be a measure of the rotatory power of the fluid.

Another method is as follows:—Use a single quartz instead of a biquartz; in front of it place a pair of quartz wedges, with the thin end of one opposite the thick end of the other; the outer surfaces having been cut perpendicularly to the axis. If the plate be right-handed, the wedges must be left-handed, and *vice versâ*. The wedges must be made to slide one over another so as together to form a plate of any required thickness, and a scale connected with the sliding gear registers the thickness of the plate produced. When the tube is removed, the wedges are adjusted so as to compensate the quartz plate, and their position is considered as the zero point of the scale. When the tube is replaced, the wedges are again adjusted so as to compensate the action of the fluid in the tube, and the difference of the readings gives the thickness of quartz necessary for the compensation. The rotatory effect of a given thickness of quartz being supposed known, we know at once the effect of a thickness of the fluid under examination equal to the length of the tube.

Another method has been based upon the principle of Savart's bands (described below); but sufficient has perhaps here been said to illustrate the principle of the saccharometer.

Circular polarisation may, however, be also produced by other means, namely, by total reflexion, and by transmission through doubly-refracting plates of suitable thickness.

It will, perhaps, be best to begin with the last. And, in order the better to understand the process, we must consider briefly the result of compounding two rectilinear vibrations under different circumstances.

Suppose a particle of ether to be disturbed from its point of rest O in a direction O A. The attraction of the particles in its neighbourhood would tend to draw it back to O; and let O A be the extreme distance to which under these attractions it would move.

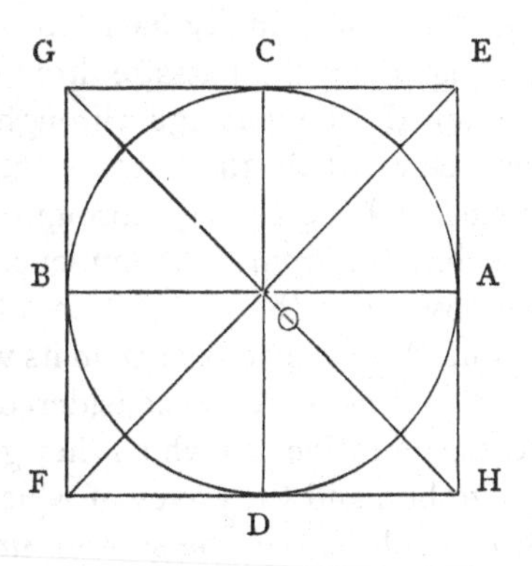

Having reached A, it would return to O; and passing through O with a velocity equal to that with which it started under the disturbing force, it would move to a point B equidistant from O with A, but in the opposite direction. And if, as is generally supposed, the ether is perfectly elastic, or that there are no internal frictions or other conditions whereby the energy of motion is converted into other forms of energy, the oscillations or vibrations of the particle between the points A and B will continue indefinitely. Now, suppose that while these vibrations are going on, a second disturbing impulse equal in intensity, but in a direction at right angles to the first, be communicated to the particle; it is clear that the effect on the motion of the particle will be different according as it takes place at the point of greatest velocity O, or at that of no velocity A or B, or at some intermediate point.

Our object is to consider the effects under these various circumstances.

A complete vibration consists in the motion from O to A, thence to B, and finally back to O; so that if O be the starting point, the passage through A will be removed one-fourth, the passage through O from A towards B will be one-half, the passage through B will be three-fourths, and the passage through O from B to A a complete vibration from the commencement. This being so, suppose that the second impulse be communicated while the particle is at O on its way towards A, then the impulses may be considered as simultaneous, and the vibrations to which they give rise will commence together, and the waves of which they form part will be coincident. If the second impulse take place when the particle is at A, the two sets of vibrations or waves to which they belong will have a difference of phase (*i.e.* the first will be in advance of the second) equal to one-fourth of a vibration or one-fourth of a wave length. If the second impulse take place when the particle is at O on its way to B, the difference of phase will be half; if when it is at B, the difference will be three-fourths of a wave length.

The particle, being at O, and subject to two simultaneous impulses of equal strength, one in the direction of A, the other in that of C, must move as much in the direction of C as in that of A; that is, it must move in a straight line equally inclined to both, namely, O E in the same figure. And inasmuch as the two impulses in no way impede one another, the particle will move in each direction as far as it would have done if the other had not taken place. In other words, if we draw a square about O, with its sides at a distance equal to O A or O B, the extent of the vibration will be represented by O E, where E is a corner of the square. The complete vibration will then be represented by the diagonal E F, in the same way as it was by the line A B in the first instance. If the impulse had been communicated at the instant of passage through O on the way to B, it is clear that a similar train of reasoning would have shown that the vibration would have been in the other diagonal G H. We conclude, therefore, that if two sets of rectilinear vibrations, or plane waves, at right angles to one another, combine, then when they are coincident they will produce a rectilinear vibration, or wave, whose plane is equally inclined to the two, and lying in the direction towards which the motions are simultaneously directed. In the figure this is represented by the dexter diagonal. When the two sets of waves have a difference of phase equal to half a wave length, their combination gives rise to a wave represented in the figure by the sinister diagonal.

Suppose now that the second impulse is communicated at the instant when the particle is at A; in other words, that the two sets of waves have a difference of phase equal to one-fourth of a wave length. At that instant the particle will have no velocity in the direction of A B (for convenience, say eastwards), and will consequently begin to move in the direction of the second impulse, say northwards. But as time goes on, the particle will have an increasing velocity westwards and a diminishing velocity northwards; it will therefore move in a curve which gradually and uniformly bends, until, when it has reached its greatest distance northwards, it will be moving wholly westwards. And as the motion not only will be the same in each quadrant, but would be the same even if the directions of the impulses were reversed, it is clear that the curvature of the path will be the same throughout; that is to say, if two sets of waves of the same magnitude in planes perpendicular to one another, and with a difference of phase equal to one-fourth of a wave length combine, they will produce a wave with circular vibrations.

If the second impulse be given when the particle arrives at B—that is, if the waves have a difference of phase equal to three-fourths of a wave length—similar considerations will show that the motion will be circular, but in the opposite direction.

Suppose, therefore, that we allow plane-polarised light to fall upon a plate of doubly refracting crystal cut perpendicularly to the axis in the case of a uni-axal crystal, or in the case of a bi-axal to the plane containing the two axes—say a plate of mica, which splits easily in that direction—then the vibrations will, as before explained, be resolved in two directions, at right angles to one another. And further, if the original directions of vibration be equally inclined to the new directions, *i.e.* if it be inclined at 45° to them, the amount or extent of vibration resolved in each direction will be equal. Further, if the thickness of the plate be such as to produce retardation or difference of phase equal to a quarter of a wave, or an odd number of quarter-wave lengths, for the particular ray under consideration, then the two sets of vibrations on emerging from the mica plate will recombine, and, in accordance with the reasoning given above, they will form a circular vibration, left-handed or right-handed according as the retardation amounts to an integral number of three-quarter wave lengths or not.

It thus appears that a plate of mica, or quarter undulation plate, which retards one of the sets of waves into which it divides an incident set by an odd multiple of quarter-wave lengths, affords a means of producing circular from plane polarisation. It remains to be shown that, with the same plate in different positions, right or left-handed circular polarisation may be produced at pleasure. Suppose that the original vibrations are in the direction E F in the foregoing figure; the mica plate will resolve them into the two directions A B, C D, one of the rays—say the first—will be transmitted with greater velocity than the other, and the vibrations along C D will be one-fourth of a wave length behind those along A B. This will correspond to the case discussed above, and will give rise to a circular vibration in a direction opposite to that of the hands of a clock. Suppose, however, that the plate be turned round through a right angle, so that the vibrations which are transmitted with greater velocity are placed parallel to C D, and those which are transmitted with lesser along A B. The ray whose vibrations are along A B will then be a quarter-wave length in advance, or, what comes to the same thing, they are three-quarters of a wave length in rear of the others; and this condition of things produces, as explained before, a circular vibration in a direction the reverse of the former. It thus appears that the plate placed in one direction will convert plane into right-handed circular polarisation; and if turned round through a right angle from that position will convert plane into left-handed circular polarisation. A like change from right-handed to left-handed circular polarisation, or *vice versâ*, may obviously be effected by turning the original plane of polarisation through a right angle; so that it shall lie between lines of concurrent instead of between lines of discordant motion.

It must be borne in mind that the above remarks refer to light of one colour only, because the retardation due to the thickness of the plate will amount exactly to a half-wave of one definite length only. But if we take a thickness calculated to cause such a retardation in a yellow ray, at or near one of the sodium lines, the plate will be found suitable for general approximate use with all luminous rays. This observation, of course, will apply, *mutatis mutandis*, to all experiments with quarter-undulation plates.

CHAPTER V.

CIRCULAR POLARISATION BY REFLEXION.

THE conversion of plane into circularly polarised light may also be effected by total reflexion. If plane-polarised light traversing glass be incident upon the inner side of the limiting surface at any angle at which total reflexion takes place, it may be considered as resolved into two plane-polarised rays, the vibrations of one being parallel and those of the other perpendicular to the plane of reflexion; and there is reason to believe that in every such case a difference of phase is brought about which for a particular angle in each substance (in St. Gobain glass it is 54° 30′) it has a maximum value somewhat in excess of one-eighth of a wave length. There are two angles, one on each side of that giving the maximum, for which the difference of phase is one-eighth of a wave length. And if the original plane of vibration be inclined at an angle of 45° to that of reflexion, the amplitudes of the two vibrations, into which the reflected vibrations are supposed to be resolved, will be equal. A full discussion of the mechanical causes which may be considered to effect this difference of phase would carry us deeper into the more difficult parts of the Wave Theory than would be suitable in this place. But if we accept the fact that the above-mentioned effects result, when polarised light (whose plane of vibration is inclined at 45° to that of reflexion, so as to ensure an equality of amplitudes in the components) is reflected at a proper angle, then the following construction will be readily understood. Take a rhomb of glass, *a, b, c, d,* Fig. 14, whose acute angles are 54° 30′; a ray incident perpendicularly to either end will undergo two total internal reflexions at the sides—say at *p* and *s*—and will emerge

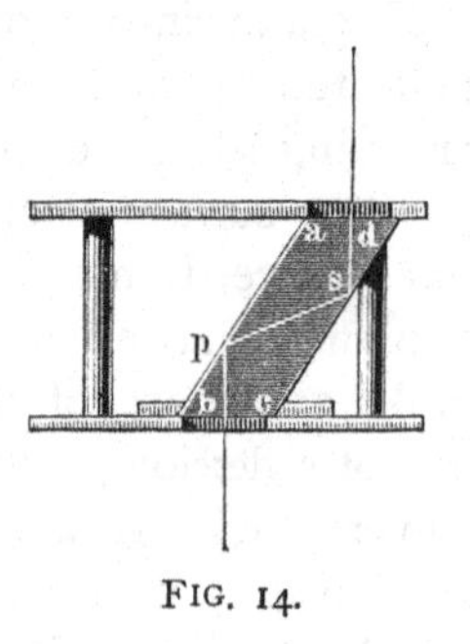

FIG. 14.

perpendicularly to the other end. These two reflexions will together produce a retardation, as described above, of one-fourth of a wave length. And if the ray be originally polarised, and its plane of vibration be inclined at an angle of 45° to that of reflexion (that of the paper in the figure), the amplitudes of the two vibrations will be equal; and all the conditions will be fulfilled for the production of circular polarisation. Such an instrument was invented by Fresnel, and is called in consequence Fresnel's rhomb. On account of its length and its displacement of the ray, it is not so convenient as a quarter-undulation plate; but, on the other hand, it affects rays of all wave lengths equally, while the quarter-undulation plate can strictly be adapted to rays of only one wave length.

If either of these instruments be introduced and suitably placed between a selenite plate and the analyser, the chromatic effects will be similar to those due to a plate of quartz cut perpendicularly to the axis.

Another important property of these instruments consists in their effect upon circularly polarised light. Such light may be considered to arise from two plane-polarised rays whose vibrations are perpendicular to one another, and which present a difference of phase equal to a quarter of a wave length. If, therefore, either a quarter-undulation plate or a Fresnel's rhomb be suitably placed, it will either increase or diminish the difference of phase by a quarter of a wave length. In the one case the difference of phase will amount to a half-wave length, in the other it will vanish. And in either case the vibration will be converted into a rectilinear one; but the directions of vibration in the two cases will be perpendicular to one another.

Some years ago Sir Charles Wheatstone devised a very beautiful application of reflexion from a metallic surface for converting plane into circular polarisation. The following is a sketch of the principle upon which it is based. If a ray of plane-polarised light fall upon a metallic reflector, it is divided into two, whose vibrations are respectively parallel and perpendicular to the reflector; and the latter is retarded behind the former by a difference of phase depending upon the angle of

FIG. 15.

incidence. If the plane of vibration of the incident ray be inclined to the plane of incidence at an angle (nearly 45°) which varies with the metal employed, but which is perfectly definite, the intensities become equal. And further, if the angle of incidence have a particular value, dependent upon the nature of the metal (for silver 72°), the retardation will amount to a quarter of a wave length. And the result will be a circularly polarised ray, as in the case of total reflexion.

To give practical means for the use of this principle, the author of it has so modified the construction of Norremberg's polariscope as to admit of the introduction of a silvered plate in a proper position. The following description is quoted from his paper on the subject. *See* Fig. 15.

"A plate of black glass, G, is fixed at an angle of 3° to the horizon. The film to be examined is to be placed on a diaphragm, D, so that the light reflected at the polarising angle from the glass plate shall pass through it at right angles, and, after reflexion at an angle of 18° from the surface of a polished silver plate S, shall proceed vertically upwards. N is a Nicol's prism, or any other analyser, placed in the path of the second reflection. The diaphragm is furnished with a ring, movable in its own plane, by which the crystalised plate to be examined may be placed in any azimuth. C is a small movable stand, by means of which the film to be examined may be placed in any azimuth and at any inclination; for the usual experiments this is removed.

"If a lamina of quartz cut parallel to the axis, and sufficiently thin to show the colours of polarised light, be placed upon the diaphragm so that its principal section (*i.e.* the section containing the axis) shall be 45° to the *left* of the plane of reflexion, on turning the analyser from left to right, instead of the alternation of two complementary colours at each quadrant, which appear in the ordinary polarising apparatus, the phenomena of successive polarisation, exactly similar to those exhibited in the ordinary apparatus by a plate of quartz cut perpendicularly to the axis, will be exhibited; the colours follow in the order R, O, Y, G, B, P, V, or, in other words, ascend as in the case of a right-handed plate of quartz cut perpendicularly to the axis. If the lamina be now either inverted, or turned in its own plane 90°, so that the principal section shall be 45° to the right of the plane of reflexion, the succession of the colours will be reversed, while the analyser moves in the same direction as before, presenting the same phenomena as a left-handed plate of quartz cut perpendicularly to the axis. Quartz is a positive doubly refracting crystal; and in it consequently the ordinary index of refraction is smaller than the extraordinary index. But if we take lamina of a negative crystal, in which the extraordinary index is the least, as a film of Iceland spar split parallel to one of its natural cleavages, the phenomena are the reverse of those exhibited by quartz: when the principal section is on the *left* of the plane of reflexion the colours descend, and when it is on the *right* of the same plane the colours ascend, the analyser being turned from left to right.

"It has been determined that the ordinary ray, both in positive and negative crystals, is polarised in the principal section, while the extraordinary ray is polarised in the section perpendicular thereto. It is also established that the index of refraction is inversely as the velocity of transmission. It follows from the above experimental results, therefore, that when the resolved ray whose plane of polarisation is to the left of the plane of reflexion is the quickest, the successive polarisation is right-handed, and when it is the slowest, the successive polarisation is left-handed—in the order R, O, Y, G, B, P, V, and in the second case in the reversed order.

"The rule thus determined is equally applicable to laminæ of bi-axal crystals.

"As selenite (sulphate of lime) is an easily procurable crystal, and readily cleavable into thin laminæ capable of showing the colours of polarised light, it is most frequently employed in experiments on chromatic polarisation. The laminæ into which this substance most readily splits contain in their planes the two optic axes; polarised light transmitted through such laminæ is resolved in two rectangulrr directions, which respectively bisect the angles formed by the two optic axes: the line which bisects the smallest angle is called the intermediate section; and the line perpendicular thereto which bisects the supplementary angle is called the supplementary section. These definitions being premised, if a film of selenite is placed on the diaphragm with its intermediate section to the left of the plane of reflexion, the successive polarisation is direct or right-handed; if, on the contrary, it is placed to the right of that plane, the successive polarisation is left-handed. The ray polarised in the intermediate section is therefore the most retarded; and as that section is considered to be equivalent to a single optic axis, the crystal is positive.

"In one kind of mica the optic axes are in a plane perpendicular to the laminæ. They are inclined $22\frac{1}{2}°$ on each side of the perpendicular within the crystal, but, owing to the refraction, are seen respectively at an angle of $35°·3$ therefrom. The principal section is that which contains the two optic axes. If the film is placed on the diaphragm with its principal section inclined 45° to the left of the plane of reflexion, the successive polarisation is right-handed. The ray, therefore, polarised in the section which contains the optic axes is the one transmitted with the greatest velocity.

"Films of uni-axal crystals, whether positive or negative, and of bi-axal crystals, all agree therefore in this respect:—that if the plane of polarisation of the quickest ray is to the left of the plane of reflexion, the successive polarisation is right-handed when the analyser moves from left to right; and if it is to the right of the plane of reflexion, other circumstances remaining the same, the successive polarisation is left-handed.

"It must be taken into consideration that the principal section of the film is inverted in the reflected image; so that if the plane of polarisation of the quickest ray in the film is to the left of the plane of reflexion, it is to the right of that plane in the reflected image.

"It may not be uninteresting to state a few obvious consequences of this successive polarisation in doubly refracting laminæ, right-handed and left-handed according to the position of the plane of polarisation of the quickest ray. They are very striking as experimental results, and will serve to impress the facts more vividly on the memory.

"1. A film of uniform thickness being placed on the diaphragm with its principal section 45° on either side of the plane of reflexion, when the analyser is at 0° or 90° the colour of the film remains unchanged, whether the film be turned in its own plane 90°, or be turned over so that the back shall become the front surface; but if the analyser be fixed at 45°, 135°, 225°, or 315°, complementary colours will appear when the film is inverted from back to front, or rotated in its own plane either way 90°.

"2. If a uniform film be cut across and the divided portions be again placed together, after inverting one of them, a compound film is formed, which, when placed on the diaphragm, exhibits simultaneously both right-handed and left-handed successive polarisation. When the analyser is at 0° or 90° the colour of the entire film is uniform; as it is turned round the tints of one portion ascend, while those of the other descend; and when the analyser is at 45° or $n90° + 45°$, they exhibit complementary colours.

"3. A film increasing in thickness from one edge to the other is well suited to exhibit at one glance the phenomena due to films of various thicknesses. It is well known that such a film placed between a polariser and an analyser will show, when the two planes are parallel or perpendicular to each other and the principal section of the film is intermediate to these two planes, a series of parallel coloured bands, the order of the colours in each band from the thick towards the thin edge being that of their refrangibilities, or R, O, Y, G, B, P, V. The bands seen when the planes are perpendicular are intermediate in position to those seen when the planes are parallel; on turning round the analyser these two systems of bands alternately appear at each quadrant, while in the intermediate positions they entirely disappear.

"Now let us attend to the appearances of these bands when the wedge-form film is placed on the diaphragm of the instrument, Fig. 15. As the analyser is moved round, the bands advance toward or recede from the thin edge of the wedge without any changes occurring in the colours or intensity of the light, the same tint occupying the same place at every half-revolution of the analyser. If the bands advance toward the thin edge of the wedge, the successive polarisation of each point is left-handed; and if they recede from it, the succession of colours is right-handed; every circumstance, therefore, that with respect to a uniform film changes right-handed into left-handed successive polarisation, in a wedge of the same substance transforms receding into advancing bands, and *vice versâ*. These phenomena are also beautifully shown by concave or convex films of selenite or rock-crystal, which exhibit concentric rings contracting or expanding in accordance with the conditions previously explained.

"4. Few experiments in physical optics are so beautiful and striking as the elegant pictures formed by cementing laminæ of selenite of different thicknesses (varying from $\frac{1}{2000}$ to $\frac{1}{50}$ of an inch) between two plates of glass. Invisible under ordinary circumstances, they exhibit, when examined in the usual polarising-apparatus, the most brilliant colours, which are complementary to each other in the two rectangular positions of the analyser. Regarded in the instrument, Fig. 13, the appearances are still more beautiful; for, instead of a single transition, each colour in the picture is successfully replaced by every other colour. In preparing such pictures it is necessary to pay attention to the direction of the principal section of each lamina when different pieces of the same thickness are to be combined together to form a surface having the same uniform tint; otherwise in the intermediate transitions the colours will be irregularly disposed.

"5. A plate of rock-crystal cut perpendicular to the axis loses its successive polarisation, and behaves exactly as an ordinary crystallised film through which rectilinear polarised light is transmitted.

"By means of the phenomena of successive polarisation it is easy to determine which is the thicker of two films of the same crystalline substance. Place one of the films on the diaphragm E of the instrument (Fig. 15) in the position to show, say, right-handed polarisation; then cross it with the other film; if the former be the thicker, the successive polarisation will be still right-handed; if both be equal, there will be no polarisation; and if the cross film be the thicker, the successive polarisation will be left-handed. In this manner a series of films may be readily arranged in their proper order in the scale of tints.

"In the experiments I have previously described the planes of reflexion of the polarising mirror and of the silver plate were coincident; some of the results obtained when the azimuth of the plane of reflexion of the silver plate is changed are interesting.

"I will confine my attention here to what takes place when the plane of reflexion of the silver plate is 45° from that of the polarising reflector.

"When the principal sections of the film are parallel and perpendicular to the plane of reflexion of the polarising mirror, as the whole of the polarised light passes through one of the sections, no interference can take place, and no colour will be seen, whatever be the position of the analyser.

"When the principal sections of the film are parallel and perpendicular to the plane of reflexion on the silver plate, they are 45° from the plane of reflexion of the polarising mirror.

"The polarised ray is then resolved into two components polarised at right angles to each other; one component polarised in the plane of reflexion of the silver plate, the other perpendicular thereto; and one is retarded upon the other by a quarter of an undulation.

"When the analyser is 0° or 90° no colours are seen, because there is no interference; but when it is placed at 45° or 135°, interference takes place, and the same colour is seen as if light circularly polarised had been passed through the film. The bisected and inverted film shows simultaneously the two complementary colours.

"But when the film is placed with one of its principal sections 22½° from the plane of reflexion of the polarising mirror, on turning round the analyser the appearances of successive polarisation are reproduced exactly as when the planes of reflexion of the silver plate and of the polarising mirror coincide. In this case the components of the light oppositely polarised in the two sections are unequal, being as cos $22\frac{1}{2}°$ to sin $22\frac{1}{2}°$; these components respectively fall $22\frac{1}{2}°$ from the plane of reflexion of the silver plate and from the perpendicular plane, and are each resolved in the same proportion in these two planes. The weak component of the first, and the strong component of the second, are resolved into the normal plane, while the strong component of the first and the weak component of the second are resolved into the perpendicular plane.

"The apparatus (Fig. 15) affords also the means of obtaining large surfaces of uncoloured or coloured light in every state of polarisation—rectilinear, elliptical, or circular.

"It is for this purpose much more convenient than a Fresnel's rhomb, with which but a very small field of view can be obtained. It must, however, be borne in mind that the circular and elliptical undulations are inverted in the two methods: in the former case they undergo only a single, in the latter case a double reflexion.

"For the experiments which follow, the crystallised plate must be placed on the diaphragm E, between the silver plate and the analyser, instead of, as in the preceding experiments, between the polariser and the silver plate.

"By means of a moving ring within the graduate circle D the silver plate is caused to turn round the reflected ray, so that, while the plane of polarisation of the ray remains always in the plane of reflexion of the glass plate, it may assume every azimuthal position with respect to the plane of reflexion of the silver plate. The film to be examined and the analyser move consentaneously with the silver plate, while the polarising mirror remains fixed.

"In the normal position of the instrument the ray polarised by the mirror is reflected unaltered by the silver plate; but when the ring is turned to 45°, 135°, 225°, or 315°, the plane of polarisation of the ray falls 45° on one side of the plane of reflexion of the silver plate, and the ray is resolved into two others, polarised respectively in the plane of reflexion and the perpendicular plane, one of which is retarded on the other by a quarter of an undulation, and consequently gives rise to a circular ray, which is right-handed or left-handed according to whether the ring is turned 45° and 225°, or 135° and 315°. When the ring is turned so as to place the plane of polarisation in any intermediate position between those producing rectilinear and circular light, elliptical light is obtained, on account of the unequal resolution of the ray into its two rectangular components.

"Turning the ring of the graduated diaphragm from left to right, when the crystallised film is between the silver plate and the analyser, occasions the same succession of colours for the same angular rotation as rotating the analyser from right to left when the instrument is in its normal position and the film is between the polariser and the silver plate."

The same principles apply to the case of bi-axal crystals cut parallel to a plane containing the two optic axes. A ray of plane-polarised light transmitted through such a plate is divided into two, whose vibrations respectively bisect the angles formed by the two axes. As mentioned above, the line which bisects the smallest angle is called the intermediate section, and the line perpendicular to it the supplementary section; and the order of the colours depends upon the relative velocity of the two rays. In selenite, the ray whose vibrations lie in the supplementary section is the slowest; in mica it is the swiftest. Hence these two crystals, all other circumstances being alike, give the colours in opposite orders, and may be regarded as positive and negative, like quartz and Iceland spar. And a test similar to that indicated for uni-axal may be applied to bi-a al crystals.

Some interesting and varied experiments may be made by using two circularly polarising instruments, *e.g.* two quarter-undulation plates (say the plates A and B), one between the polariser proper and the crystal (C) under examination, the other between the crystal and the analyser. The light then undergoes the following processes. If the plate A be placed so that its axis is at 45° on one side or other of the original plane of vibration, and the plate B with its axis perpendicular to that of A, then on turning the analyser we shall have the phenomena of circular polarisation described above. Again, if, the plates A and B retaining the positions before indicated, the crystal C be turned round in its own plane, then, since the light emerging from A and B is circularly polarised, it has lost all trace of direction with reference to the positions of polariser and analyser, and consequently no change will be observed.

Again, if the plates A and B have their axes

directed at 45° on either side of the axis of C, and the three plates be turned round as one piece, the colour will remain unchanged, while if the analyser be turned the colours will follow in the regular order. If the plates A and B have their axes directed at 45° on the same side of the axis of C, and the pieces be turned round bodily as before, the colours change in the same order as above, and go through their cycle once in every right angle of rotation; and if the analyser be turned in the same direction, the colours change, but in the reverse order. The explanation of this is to be found in the fact that when the plates A and B are crossed, the retardation due to A is compensated by that due to B; so that the only effective retardation is that due to the crystal C. But upon the latter depends the rotation of the plane of vibration; if, therefore, the polariser and analyser remain fixed, the colour will remain unaltered. When the plates A and B have their axes parallel, there is no compensation, and the colour will consequently change. It should be added that the rotation of the plane of vibration, and consequently the sequence of colours, does not follow exactly the same law in these cases as in quartz.

Mention was made above of the bands produced in the spectra of polarised light. Beside the fact of the existence of these bands, it has been found upon examination that the state of polarisation at different parts of the interval between two successive bands varies; and such an examination may be made by means of a quarter-undulation plate or a Fresnel's rhomb.

If we carefully examine the spectrum of light which has passed through a selenite, or other ordinary crystal, placed with its axis at 45° to the plane of vibration, we shall find on turning the analyser that, commencing with two consecutive bands in position, the parts occupied by the bands and those midway between them are plane-polarised, for they become alternately dark and bright; while the intermediate parts, *i.e.* the parts at one-fourth of the distance from one band to the next, remain permanently bright. These are, in fact, circularly polarised. But it would be incorrect to conclude from this experiment alone that such is really the case, because the same appearance would be seen if those parts were unpolarised, *i.e.* in the condition of ordinary light. And on such a supposition we should conclude, with equal justice, that the parts on either side of the parts last mentioned (*i.e.* the parts separated by one-eighth of the interval between two bands) were partially polarised. But if we introduce a quarter-undulation plate between the selenite and analyser, with its axis inclined at 45° to that of the selenite, circular polarisation will be converted into plane and plane into circular. This being so, the parts which were originally banded ought to become bright and to remain bright, while those that were originally bright ought to become banded during the rotation of the analyser. The effect to the eye will consequently be a general shifting of the bands through one-fourth of the space which separates each pair. Further, as on the one hand plane polarisation is converted into circular right-handed or left-handed by two positions of the plate at right angles to one another, so on the other right-handed circular polarisation will be converted by the plate in a given position into plane polarisation having the vibrations in one direction, and left-handed into plane polarisation having the vibrations in a direction at right angles to the former. Hence, if the plate be turned through a right angle from the position first described, the bands will be shifted in a direction opposite to that in which they were moved at first. In this we have evidence not only that the polarisation on either hand is circular, but also that on the one side it is right-handed, while on the other it is left-handed.

CHAPTER VI.

PHENOMENA PRODUCED BY MECHANICAL MEANS—UNANNEALED GLASS.

ALL the phenomena hitherto described manifestly depend upon the internal structure of the crystal plate, in virtue of which it affects the vibratory movement of the ether within it differently in different directions. And seeing that most crystals, when broken, divide themselves naturally into smaller crystals having the same form, *i.e.* having their planes and edges similarly inclined, we are naturally led to conclude that the structure of these bodies may differ not so much in different parts, as along different lines or planes connected with the forms into which they break, or (as it is also described) with their planes of natural cleavage. And this suggests the question whether an uncrystalline body might not, by pressure, or strain, or other mechanical distortion, be caused to affect the motions of the ether within it in a manner dependent upon their direction, and in that way to exhibit chromatic effects with polarised light analogous to those described above. Experiment answers this question in the affirmative.

The simplest experiment in this branch of inquiry consists in taking a rectangular bar of ordinary glass; and having crossed the polariser and analyser so as to give a dark field, to strain the bar with both hands as if we were trying to bend it or to break it across. The side towards which it may be supposed to be bent is of course compressed, while the opposite is stretched out. Between these two there must be an intermediate band, more or less midway between the two, which is neither compressed nor stretched. The moment the strain is put upon the bar light will be seen to pass through the parts of the bar nearest to both sides, while a band remains dark midway between the two.

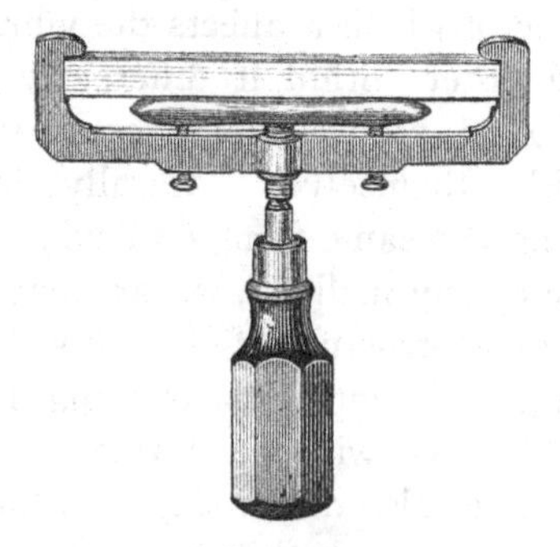

FIG. 16.

This shows that the mechanical strain has imparted to portions of the glass a structural character analogous, at all events optically, to that of a crystal. The effects may be increased and rendered more striking by placing the glass in a frame furnished with a screw, by which the rod may be firmly held and considerable pressure applied at particular points (Fig. 16). When this is done the structural character becomes more completely developed, and the dark band is fringed with colours which appear to flow inwards or outwards according as the pressure is increased or diminished. A slightly different, but more effective, exhibition of chromatic polarisation is produced by squeezing a thick square plate of glass in a vice. In this case the pressure may be carried further without fear of fracture, and the chromatic effects heightened (Fig. 17).

FIG. 17

It is, however, well known that molecular forces, such as those due to heat and cooling, in many cases far transcend in intensity those which we can exert by mechanical arrangements. And, in fact, if a block of glass be unequally heated to a very moderate degree, the internal structural effects immediately reveal themselves by dark bands, which serve as indications of strain and pressure. As the block cools, these landmarks gradually disappear, and the field becomes again uniformly dark. But by far the most splendid effects (and these are permanent) are produced by unannealed glass; that is, by glass which has been rapidly and therefore unequally cooled. When a mass of glass has been cast in a mould in the form of a thick plate, then, whatever be the contour line, the outside will cool first and become a rigid framework to which the interior of the mass must accommodate itself. The nature and direction of the pressure at each point of the interior will be primarily dependent upon the form of the contour; and by adopting various forms of contour the most beautiful and varied figures with coloured compartments may be produced. Among the dark bands some will be found to retain their position while the glass is turned round; these answer to the systems of isochromatic curves in uni-axal and bi-axal crystals, and mark the borders between regions of strain and of pressure. Beside these there are also variable bands, which correspond to the brushes which cross the isochromatic curves. The forms and colours of the figures produced by transparent bodies when submitted to polarised light have been conversely used as a means of measuring, with almost unparalleled accuracy, the mechanical pressures which such a body is undergoing.

CHAPTER VII.

ATMOSPHERIC AND OTHER POLARISATION—THE POLAR CLOCK.

WE may here digress with advantage for a few moments from the direct course of our subject, in order to notice some instances of the phenomena of polarisation which occur in everyday experience, as distinguished from laboratory experiments and instrumental contrivances. But the reader will doubtless remember that to these, as well as to all other cases, the remark made at the outset applies, viz. that instrumental aid must be invoked in order to render them perceptible by our eyes.

Besides glass, many other substances may be used as reflectors so as to produce polarisation; for example, leaves of trees, particularly ivy, mahogany furniture, windows, shutters, and often roofs of houses, oil paintings, &c., and last, but not least, the surface of water. In each of these cases, when the reflected beam is examined with a Nicol, the alternations of light and darkness are most strongly marked, and the colours (if a crystal plate be used) are most vivid, or in technical language the polarisation is most complete, when the light is reflected at a particular angle. In proportion as the inclination of the incident light deviates from this angle, the colours become fainter, until, when it deviates very greatly, all trace of polarisation disappears.

The minor practical applications of these facts are numerous, and several may suggest themselves to the reader, especially if he carries a Nicol and a quartz plate in his pocket. Among them the following may be mentioned:—A Nicol so placed as to extinguish superficial reflexion or glare will often be found of considerable use in examining oil pictures placed inconveniently in respect of light. Again, it has been suggested that a similar contrivance might be used by persons engaged in the sport of fishing, in order the better to see the fish or other objects at a moderate depth below the surface of the water. A Nicol's prism has also been placed with very good effect in the eyepiece of a telescope (the precise disposition depending upon the internal construction of the instrument), in order to diminish the intensity of sunlight. This arrangement has the additional advantage of reducing the heat also, because the rays of heat are affected in respect of polarisation very nearly in the same way as those of light.

It will be found very interesting to examine the polarisation of sunshine reflected from ripples on the surface of a lake, or better still from the waves of the sea, and its different degrees of completeness produced at the variously inclined portions of the waves. But, without having recourse to nature on so large a scale, an artificial piece of water may be placed in our room.

A tea-tray will serve as well as anything else to form our little sea; and a periodic tap at one corner will cause ripple enough for the present purpose. The waves appear bright, and although brighter in some parts than in others, they are nowhere entirely dark. But on turning the Nicol round, the contrast of light and darkness becomes much stronger than before. In parts the light is absolutely extinguished, or the polarisation is complete; in others it is incomplete in various degrees. And if a selenite or other crystal plate be introduced, we have the beautiful phenomena of iris-coloured rings playing over the surface of our miniature sea.

Suppose that we now turn our attention to the sky, and on a clear bright day we sweep the heavens with a polariscope, or even with a mere Nicol's prism; we shall find traces of polarisation in many directions. But if we observe more closely we shall find that the most marked effects are produced in directions at right angles to a line drawn from our eye to the sun—when, in fact, we are looking across the direction of the solar beams. Thus, if the sun were just rising in the east or setting in the west, the line of most vivid effect would lie on a circle traced over the heavens from north to south. If the sun were in the zenith, or immediately overhead, the most vivid effects would be found on the horizon; while at intermediate hours the circle of strongest polarisation would shift round at the same rate as the shadow on a sun dial, so as always to retain its direction at right angles to that of a line joining ourselves and the sun.

Now, what is it that can produce this effect, or,

indeed, what produces the effect of light from all parts of a clear sky? The sky is pure space, with no contents, save a few miles of atmosphere of the earth, and beyond that the impalpable ether, supposed to pervade all space, and to transmit light from the furthest limits of the stellar universe. The ether is, however, certainly inoperative in the diffusion of light now under consideration. But a very simple experiment will suffice to show that such a diffusion, or, as it has been better called, a scattering of light, is due to the presence of small particles in the air. If a beam from an electric lamp or from the sun be allowed to pass through a room, its track becomes visible by its reflexion from the motes of floating bodies—in fact, by the dust in the air. But if the air be cleared of dust by burning it with a spirit lamp placed underneath, the beam disappears from the parts so cleared, and the space becomes dark. If, therefore, the air were absolutely pure and devoid of matter foreign to it, the azure of the sky would no longer be seen and the heavens would appear black; the illumination of objects would be strong and glaring on one side, and on the other their shadows would be deep and unrelieved by the diffused light to which we are accustomed. Now, beside the dust, there are always minute particles of water floating in the atmosphere. These vary in size from the great rain drops which fall to earth on a sultry day, through intermediate forms of mist and of fine fleecy cloud, to almost invisible minuteness. It is these particles—whether of water or other matter it is perhaps difficult to say—which scatter the solar rays and suffuse the heavens with light. And it is a remarkable fact, established by Professor Tyndall, while operating with minute traces of gaseous vapours, that while coarser particles scatter rays of every colour—in other words, scatter white light—finer particles scatter fewer rays from the red end of the spectrum, while the finest scatter only those from the blue end. And in accordance with this law clouds are white, clear sky is blue.

But the point which most concerns us here is the fact, also discovered by Professor Tyndall, that light scattered laterally from fine particles is polarised. The experiment by which this is most readily shown is as follows:—Allow a beam of solar or other strong light to pass through a tube about thirty inches long, filled with water with which a few drops of mastic dissolved in alcohol have been mixed. The fluid so formed holds fine particles of mastic in a state of suspension, which scatter the light laterally; and if the scattered light be examined with a Nicol, traces of polarisation will be detected. But, better still, as suggested by Professor G. G. Stokes, instead of using the scattering particles as a polariser, and the Nicol as an analyser, we may polarise the light before it enters the tube, and use the particles as an analyser, and thus produce the same effect as before, not only upon the particular point of the beam to which the eye is directed, but upon the whole body of scattered light. As the Nicol is turned, the light seen laterally begins to fade; and when the instrument has been turned so as to cut off all vertical vibrations, the only parts remaining visible in a horizontal direction will be those reflected from the larger impurities floating in the water independently of the mastic. The direction of vibration of the light polarised by lateral scattering is easily remembered by the fact that the vibrations must be perpendicular both to the original and to the scattered beam; if, therefore, the latter be viewed horizontally, they must be perpendicular to two horizontal straight lines at right angles to one another, *i.e.* they must be vertical.

An effect still more beautiful, and at the same time, perhaps, more instructive, may be produced by interposing a plate of quartz between the Nicol and the tube. The whole beam then becomes suffused with colour, the tint of which changes for a given position of the spectator with the angle through which the Nicol is turned.

And not only so, but while the Nicol remains at rest the tints are to be seen scattered in a regular and definite order in different directions about the sides of the beam. But this radial distribution of colours may also be shown in a more striking manner by using a biquartz, which, as explained before, distributes the colours in opposite directions. The beam should in every case be viewed at right angles; the more obliquely it is viewed the less decided is the polarisation.

The colours here seen are those which would be observed upon examining a clear sky in a position 90° from that of the sun; and the exact tint visible will depend upon the position in which the Nicol is held, as well as upon that of the sun. Suppose, therefore, that a Nicol and quartz plate be directed to that part of the sky which is all day long at right angles to the sun—that is, to the region about the north pole of the heavens (accurately to the north pole at the vernal and autumnal equinox)—then, if on the other hand the Nicol be turned round, say in a direction opposite to that of the sun's motion, the colours will change in a definite order; if, on the other, the Nicol remain stationary while the sun moves round, the colours will change in a similar manner And thus in the latter case we might conclude the position of the sun, or, in other words, the time of the day, by the colours so shown. This is the principle of Sir Charles Wheatstone's polar clock, which is one of the few practical applications which this branch of polarisation has yet found.

Figs. 18 and 19 represent general forms of this instrument, described in the following passage by the inventor :—

"At the extremity of a vertical pillar is fixed, within a brass ring, a glass disc, so inclined that its plane is perpendicular to the polar axis of the earth. On the lower half of this disc is a graduated semicircle, divided into twelve parts (each of which is again subdivided into five or ten parts), and against the divisions the hours of the day are marked, commencing and terminating with VI. Within the fixed brass ring, containing the glass dial-plate, the broad end of a conical tube is so fitted that it freely moves round its own axis; this broad end is closed by another glass disc, in the centre of which is a small star or other figure, formed of thin films of selenite, exhibiting, when examined with polarised light, strongly contrasted colours; and a hand is painted in such a position as to be a prolongation of one of the principal sections of the crystalline films. At the smaller end of the conical tube a Nicol's prism is fixed so that either of its diagonals shall be 45° from the principal section of the selenite films. The instrument being

FIG. 18.—Wheatstone's Polar Clock.

FIG. 19.—Wheatstone's Polar Clock.

so fixed that the axis of the conical tube shall coincide with the polar axis of the earth, and the eye of the observer being placed to the Nicol's prism, it will be remarked that the selenite star will in general be richly coloured ; but as the tube is turned on its axis, the colours will vary in intensity, and in two positions will entirely disappear. In one of these positions a smaller circular disc in the centre of the star will be a certain colour (red, for instance), while in the other position it will exhibit the complementary colour. This effect is obtained by placing the principal section of the small central disc $22\frac{1}{2}°$ from that of the other films of selenite which form the star. The rule to ascertain the time by this instrument is as follows:— The tube must be turned round by the hand of the observer until the coloured star entirely disappears, while the disc in the centre remains red; the hand will then point accurately to the hour. The accuracy with which the solar time may be indicated by this means will depend on the exactness with which the plane of polarisation can be determined; one degree of change in the plane corresponds with four minutes of solar time.

"The instrument may be furnished with a graduated quadrant for the purpose of adapting it to any latitude; but if it be intended to be fixed in any locality, it may be permanently adjusted to the proper polar elevation, and the expense of the graduated quadrant be saved: a spirit-level will be useful to adjust it accurately. The instrument might be set to its proper azimuth by the sun's shadow at noon, or by means of a declination needle; but an observation with the instrument itself may be more readily employed for this purpose. Ascertain the true solar time by means of a good watch and a time equation table, set the hand of the polar clock to correspond thereto, and turn the vertical pillar on its axis until the colours of the selenite star entirely disappear. The instrument then will be properly adjusted.

"The advantages a polar clock possesses over a sun-dial are:—1st. The polar clock being constantly directed to the same point of the sky, there is no locality in which it cannot be employed, whereas, in order that the indications of a sun-dial should be observed during the whole day, no obstacle must exist at any time between the dial and the places of the sun, and it therefore cannot be applied in any confined situation. The polar clock is consequently applicable in places where a sun-dial would be of no avail—on the north side of a mountain or of a lofty building, for instance. 2ndly. It will continue to indicate the time after sunset and before sunrise—in fact, so long as any portion of the rays of the sun are reflected from the atmosphere. 3rdly. It will also indicate the time, but with less accuracy, when the sky is overcast, if the clouds do not exceed a certain density.

"The plane of polarisation of the north pole of the sky moves in the opposite direction to that of the hand of a watch; it is more convenient therefore to have the hours graduated on the lower semicircle, for the figures will then be read in their direct order, whereas they would be read backwards on an upper semicircle. In the southern hemisphere the upper semicircle should be employed, for the plane of polarisation of the south pole of the sky changes in the *same* direction as the hand of a watch. If both the upper and lower semicircles be graduated, the same instrument will serve equally for both hemispheres."

The following is a description of one among several

other forms of the polar clock which have been devised. This (Fig. 20), though much less accurate in its indications than the preceding, beautifully illustrates the principle.

"On a plate of glass twenty-five films of selenite of equal thickness are arranged at equal distances radially in a semicircle; they are so placed that the line bisecting the principal sections of the films shall correspond

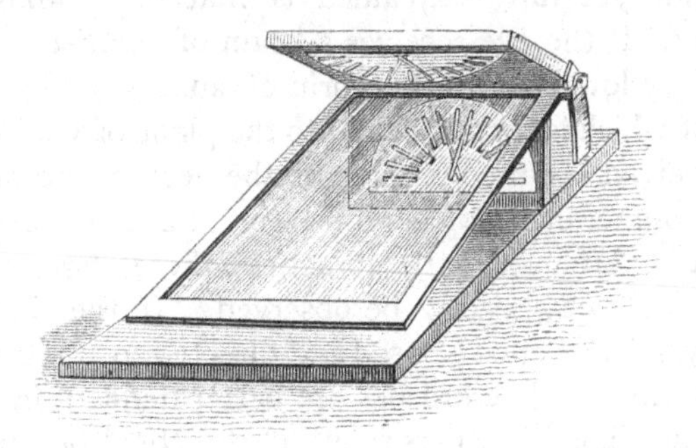

FIG. 20.—Polar Clock.

with the radii respectively, and figures corresponding to the hours are painted above each film in regular order. This plate of glass is fixed in a frame so that its plane is inclined to the horizon 38° 32′, the complement of the polar elevation; the light, passing perpendicularly through this plate, falls at the polarising angle 56° 45′ on a reflector of black glass, which is inclined 18° 13′ to the horizon. This apparatus being properly adjusted—that is, so that the glass dial-plate shall be perpendicular to the polar axis of the earth—the following will be the effects when presented towards an unclouded sky. At all times of the day the radii will appear of various shades of two complementary colours, which we will assume to be red and green, and the hour is indicated by the figure placed opposite the radius which contains the most red; the half-hour is indicated by the equality of two adjacent tints."

Among the phenomena of polarised light which may be observed either with a Nicol's prism or even with the naked eye, one of the most curious, and perhaps not yet fully explained, is that of Haidinger's brushes. If the eye receives a beam of polarised light, a pale yellow patch in the form of an hour-glass, the axis of which is perpendicular to the plane of vibration, is perceived. On either side of the neck of the figure two protuberances of a violet tint are also seen to extend. After a little practice these figures or "brushes" may readily be observed. If the day be cloudy, a Nicol must be used and directed to a tolerably bright cloud. The brushes are better defined in one position than in others; but if the Nicol be turned round, the brushes will be seen to revolve with it. If on a clear day we look in a direction at 90° from that of the sun, where the sky light is most completely polarised, the brushes may be seen with a naked eye. The reader should, however, be warned that the appearance of these brushes is not a permanent phenomenon. They are, in fact, visible for only a couple of seconds or so after the polarised light is first received on the eye. But the effect may be revived by changing the plane of polarisation. Helmholtz connects the phenomenon with some double refraction due to the yellow spots in the eye, with the area of which that of the brushes is coincident.

CHAPTER VIII.

RINGS AND BRUSHES PRODUCED BY CRYSTAL PLATES.

THE phenomena hitherto discussed are those produced by a beam of parallel rays; it remains to consider those due to systems of convergent or divergent rays. It was seen in the former case that the effect produced by a plate of crystal upon polarised light depended partly upon the direction of the ray with reference to the natural structure (or more particularly to the optic axis or axes) of the crystal, and partly to the thickness of crystal traversed by the ray. And it is clear that if a system of convergent or divergent rays fall upon a plate of crystal of uniform thickness, the rays will strike the plate at various angles, and traverse various thicknesses of the crystal. Hence it follows that the effects produced will not be, as in the case of parallel rays, fields of uniform tint, but figures of various form and tint.

It was explained above that in Iceland spar there is a particular direction, viz. that of the line joining the two opposite obtuse angles of the natural crystal, in which there is no double refraction, and in which all rays travel with the same velocity. This direction (that is to say, this line and all lines parallel to it) bears the name of the optic axis. There are many other crystals having the same property in one and only one direction—in other words, having a single optic axis. There is, moreover, another class of crystals having two such axes. Crystals of the first class, or uni-axal crystals, are again divided into two groups, viz. positive, in which the extraordinary ray is more refracted than the ordinary, and negative, in which the ordinary ray is the more refracted. It will be remembered that the ray which travels slowest is the most refracted. Among the former may be mentioned—

UNI-AXAL CRYSTALS.

Positive.

Cassiterite.	Phenakite.
Dioptase.	Quartz.
Ice.	The " Red Silver ores.
Lead Hyposulphate.	Zircon.
Magnesium Hydrate.	

Positive or Negative.

Apophyllite.	Pennine.

Negative.

Anatase.	Idocrase.
Apatite.	Mellite.
Beryl.	Mimetesite.
Calcite.	Nepheline.
Cinnabar.	Pyromorphite.
Corundum (ruby and sapphire).	Tourmaline.
	Wultenite.

Every crystal belongs to one or other of six systems or types of symmetry. These are conveniently distinguished by the various kinds of axes to which the faces are referred for the purpose of geometrical comparison.

(1.) The regular system, which is based upon a system of three equal rectangular axes. Any form derived from this will be perfectly symmetrical with reference to the three axes, and will present no distinguishing feature in relation to any of them. Crystals belonging to this system have no optic axis nor any doubly refracting property. The cube and regular octahedron are examples of forms in this system.

(2.) The tetragonal system, based upon a system of three rectangular axes, whereof two are equal, but the third greater or less than the other two. Crystals belonging to this system have one optic axis coinciding with the last-mentioned crystallographic axis. The square-based prism is a form of this system.

(3.) The hexagonal system, referred by some crystallographers to three equal axes lying in one plane, inclined at 60° to one another, and a fourth axis at right angles to the other three. Crystals of this system have one optic axis coinciding with the fourth axis above mentioned. The rhombohedron (the six faces of which are equal rhombs) is a form of this system. Iceland spar, or calcite, has cleavages parallel to the faces of a rhombohedron, the obtuse plane angles of which are 101° 5′. Some crystallographers refer the crystals of this system to axes parallel to the edges of a rhombohedron ; these axes being three in number, obliquely but equally inclined to each other, and of equal lengths.

(4.) The orthorhombic system, having three rectangular but unequal axes.

(5.) The clinorhombic system, which differs from the rhombic in this, that while one of the three axes is perpendicular to the other two, these two are oblique to one another.

(6.) The anorthic system, in which all the axes are oblique.

All crystals belonging to the last three systems have two optic axes. In the rhombic system they lie in a plane containing two of the three crystallographic axes ; in the monoclinic they lie either in the plane containing the oblique axes or in a plane, or for different colours in planes, at right angles thereto. In the triclinic no assignable relation between the optic and the crystallographic axes has been determined.

The phenomena of colours and their variations hitherto described have been produced by a beam of light, all the rays of which were parallel in their passage through the crystal or other substance under examination. There is, however, another class of phenomena due to the transmission of a convergent or divergent beam of polarised light, which we now proceed to consider.

It was shown above that the retardation due to any doubly refracting crystal, and consequently the colour produced by it, is dependent on the thickness ; and that with a crystal of constantly increasing thickness the colours go through a complete cycle, and then begin again. Suppose, then, a divergent beam to fall perpendicularly upon a uni-axal crystal plate cut at right angles to the optic axis ; the central rays will fall perpendicularly to the surface ; but the rays which form conical shells about that central ray will fall obliquely. The rays forming each shell will fall with the same degree of obliquity on different sides of the central ray, those forming the outer shells having greater obliquity than the inner. Now, the more obliquely any ray falls upon the surface the greater will be the thickness of the crystal which it traverses ; and this will still be the case even though it suffers refraction, or bending towards the perpendicular, on entering the crystal. Each incident cone of rays will consequently still form a cone when refracted within the crystal, although less divergent than at incidence, in its passage through the plate; and the successive refracted cones will be more and more oblique, as were the incident cones, but in a less degree, as we pass from the more central to the more external members of the assemblage forming the beam of light.

Now, not only will the actual thickness of crystal traversed by the ray, but so also will the doubly refracting energy, increase with the obliquity or divergence of direction from that of the axis ; and the effect of the former is in fact small compared with that of the latter in producing retardation.

Let A B C D (Fig. 21) represent the crystal plate, O P the direction of the optic axis and of the central ray, O *n*, O *n′* those of any two other rays. The ray O P will not be divided ; but O *n* will be separated by the double refraction of the plate into two, *n s*, *n r*, the one ordinary, the other extraordinary; and these will emerge parallel to one another, and may be represented by the lines *s t*, *r v*. Similarly the effect of double refraction on O n' may be represented by $n' s'$, $n' r'$, $s' t'$, $r' v'$. Suppose now that the process were reversed, and that two monochromatic rays, one ordinary, the other extraordinary, reach the plate at s and r in the directions $t s$, $v r$ respectively; these would meet at n and travel together to O. Suppose, further, that the difference in length of $s n$ and $r n$ is equal to one wave length; then, since one of them is an ordinary and the other an extraordinary ray, their vibrations will be perpendicular to one another; and if the polariser and analyser be crossed, the point n viewed

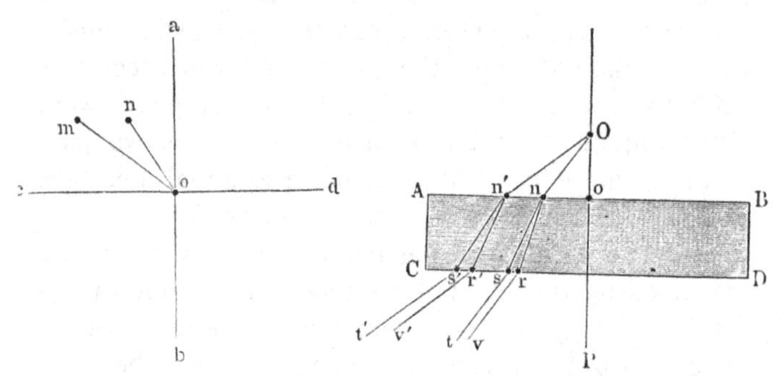

FIG. 21.

from O will appear dark. Similarly, if two rays arrive in the directions $t' s'$, $v' r'$ at the points $s' r'$ respectively, they will meet at n' and proceed together to O; and if the difference of the paths $s' n'$, $r' n'$ be two wave lengths, the point n' will also appear dark. A pair of rays reaching the crystal at points between the pairs before mentioned will emerge at a point n'' between n and n', and will present a difference of phase equal to a half-wave length. On principles explained in an earlier part of these lectures, such a point *n''* will appear bright. On either side of *n''*—that is, towards *n* and *n'*—the light will gradually fade. The same alternations of light and darkness will recur at intervals as we proceed along any straight line drawn outwards from the central ray. And inasmuch as the obliquity of the ray is the same for every point equidistant from the centre O, it follows that the phenomena of light and darkness will be the same throughout each circle drawn about the centre O. In other words, the centre will be surrounded by rings alternately bright and dark. The diameters of the rings depend, as was seen above, on the wave length of the particular light used, and will consequently be different for different coloured rays. If, therefore, white light be used, the different coloured rings would not coincide, but would be disposed in recurring series as we proceed outwards from the centre.

Another effect would, however, also be produced. Suppose the polariser and analyser to be so placed that, the field being regarded as a map, the vibrations in the one being E. and W., those in the other N. and S.; then of the two rays emerging at the most northern or the most southern point of any ring the vibrations of one would be towards the axis, or N. and S.; those of the other would be across it, or E. and W. And of these one would be extinguished by the polariser, the other by the analyser; and the same will be the case for every ring. Hence, throughout a N. and S. line crossing the entire field the light will be extinguished; and a similar effect will obviously occur along an E. and a W. line. Hence, when the polariser and analyser are crossed, the entire system of rings will be intersected by a black cross, two of whose arms are parallel to the plane of vibration of the polariser and two to that of the analyser, and the rings in the quadrants on each side of an arm are of complementary tints. When the analyser is turned round through a right angle from its former position, only one set of vibrations (say those executed in a direction E. and W.) will be extinguished, and consequently along one pair of arms of the cross the ordinary rays will pass undisturbed, along the other the extraordinary; that is to say, the cross will be white. When the polariser and analyser occupy any other position than those noticed above, there are two crosses inclined at an angle equal to that between the planes of vibration, each arm of which separates complementary rings. (*See* Plate, Figs. 1 & 2.)

Various forms of polariscopes have been devised for showing the crystal rings. The simplest of these is the tourmalin forceps, which consists of two plates of tourmalin fixed in cork discs; the latter are encircled in wire in such a way that they may be turned round in their own planes. The wire after encircling one disc is bent round so as to form a handle; it then encircles the other; and the elasticity of the wire allows the pair of discs to be opened and shut like a pair of pincers. If a crystal plate be inserted between the two, and the whole held close to the eye, the rays from parts of the field at different distances from the centre will reach the eye, having traversed the crystal with different degrees of obliquity; and a system of rings and brushes will be formed.

Another method consists in applying to Norremberg's polariscope a pair of lenses, one below the crystal with the crystal in the focus, the other above it. The first ensures that the rays shall traverse the crystal with different degrees of obliquity; the second brings within the range of vision rays which would otherwise fail to reach the eye, and at the same time converging them into a cone with a smaller vertical range, renders the ring smaller than when seen with the simple tourmalins. An additional lens of greater focal length, *i.e.* of less power, is often added in order to adjust the whole to individual eyesight.

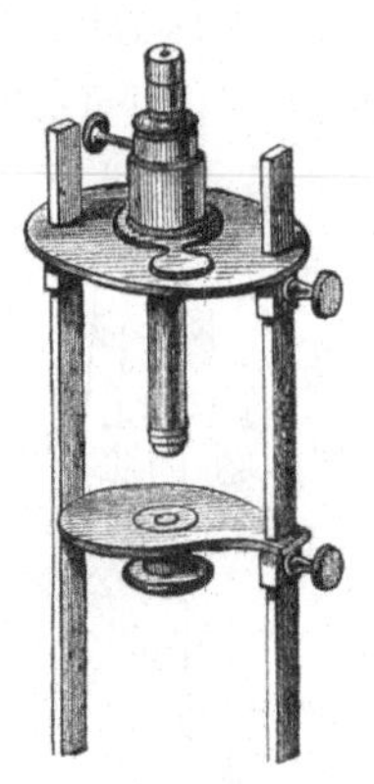

FIG. 22.

Fig. 22 gives the general appearance of the addition to the apparatus of Norremberg described above, and Fig. 23 the course of a system of rays brought to a focus on the crystal by the lens *a b*, and again converged by a second lens *c d*.

But by far the most successful arrangement for

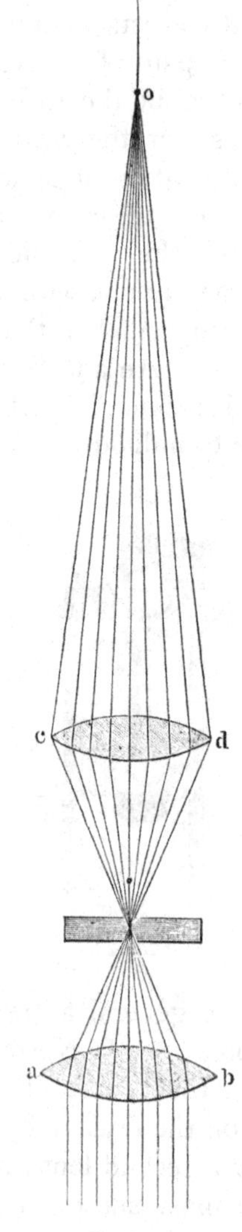

FIG. 23.

enlarging the field of view so as to comprise the complete system of rings even with bi-axal crystals having widely inclined optic axes, is the system of lenses due in the first instance to Norremberg. The disposition of the parts is shown in Fig. 24; and the general appearance of the instrument as constructed by Hofmann of Paris, and called by him the "Polarimicroscope," is also given, Fig. 25. In this instrument the lenses which converge the rays upon the crystal plate can be taken out, and replaced by others giving parallel light; it can then be used as an ordinary polariscope.

Mention has been made above of the effect of the circular polarisation of quartz in the colours produced by a beam of parallel rays of polarised light. It will be worth while to examine the modification which the rings and brushes undergo from the same cause. It has been explained that a ray of plane-polarised light in traversing a crystal of quartz in the direction of its axis is divided into two, the vibrations of which are circular, one right-handed, the other left. If the ray traverses the crystal in a direction perpendicular to the axis, and if the original vibrations are neither parallel nor perpendicular to the axis, it is also divided into two, whereof the vibrations are not circular, but rectilinear. It was suggested, first by Sir G. Airy, that these circular and rectilinear vibrations are limiting cases of elliptical; and both theory and experiment tend to confirm the suggestion, by showing that if the ray be incident on the crystal in any direction oblique to the axis, it is divided into two, the vibrations of which

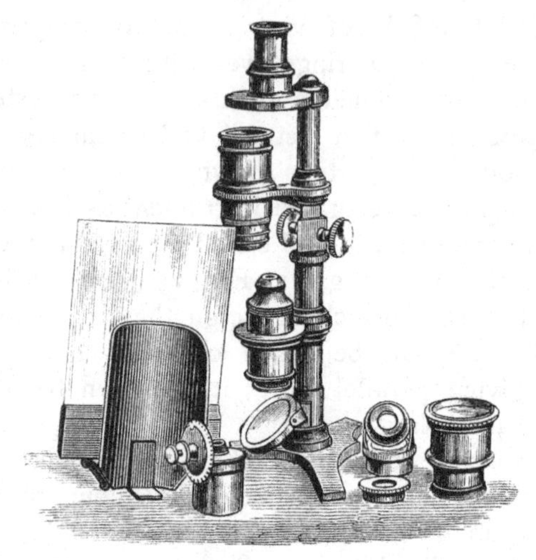

FIG. 24.

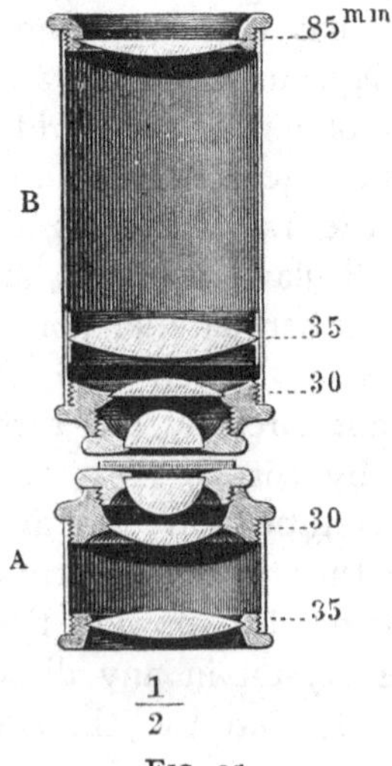

FIG. 25.

are similar ellipses having the longer diameter of the one coincident with the shorter of the other, and the motion in the two oppositely directed. The longer diameters of the ellipses coincide with the directions of vibration of the ordinary and extraordinary rays in the case of an ordinary positive crystal; and are consequently directed the one toward the centre of the figure, the other in a direction at right angles to the first.

The exact or even approximate determination of the figures produced is a complicated question, and requires mathematical analysis for its solution, but a general idea of their nature may nevertheless be easily formed. When the polariser and analyser are either parallel or crossed, circular rings are formed, and towards the outer parts of the field traces of the black cross are seen, which grow stronger as we proceed outwards from the centre—that is, towards the parts where the rays are more oblique, and where the polarisation more nearly approaches to rectilinear. But in the centre, and near to it, where the polarisation is circular or nearly so, the effects will resemble those produced by parallel rays, viz. the rays of different colours will emerge plane-polarised in different planes, and will be variously affected by the angle between polariser and analyser. In no position can they all be extinguished, and consequently in the centre all traces of the black cross will disappear.

When the planes of vibration of the polariser and the analyser are inclined at any other angle than 0° or 90°, the arms of the cross are less strongly marked, and the curvature of the rings becomes less uniform, increasing in the four points where they are crossed by the arms, and diminishing in the intermediate quadrants. When the angle between the planes of vibration is 45°, the rings assume a nearly square form, the corners of the square lying upon the lines which bisect internally and externally the angles between the planes. If the figures are produced with the analyser at 45° by two quartz plates of equal thickness, one right-handed, the other left, it will be found that the diagonals of the squares are at right angles to one another, the remains of the black cross occupying the same position in the field in both cases. (*See* Plate, Figs. 9 & 10.)

If two plates of the same thickness, the one right-handed and the other left, are placed one over the other, a beautiful effect, called from their discoverer Airy's spirals, is produced (*see* Plate, Figs. 11 & 12). In the centre of the field the rotatory powers of the plates neutralise one another, and a black cross commences. As we proceed outwards, the arms of the cross cease to be black, and become tinged with red on one side and with blue on the other. At the same time they are bent round in a spiral form, in the direction of the hands of a watch if the first plate be right-handed, and in the opposite direction if the first plate be left-handed. These spirals intersect at intervals the circular rings; the points of intersection lie in four rectangular directions, which terminate towards the outer margin of the field in four arms of a shadowy cross. The colours of the rings and spirals are more brilliant and better defined than in most other phenomena of chromatic polarisation.

A quartz plate cut parallel to the axis, when examined with convergent light, gives curves in the form of hyperbolas. These curves are wider in proportion to the thinness of the plate, but if the plate be thick enough to render the curves moderately fine, the colour becomes very faint. They may, however, be rendered distinct by using homogeneous light. The dark and light parts exchange positions when the analyser is turned through 90°. Two such plates with their axes at right angles to one another give coloured hyperbolas perfectly visible with the white light. Plates of Iceland spar exhibit similar phenomena, but the lines and curves are far more closely packed.

If the plate be cut in a direction inclined at 45° (or at any angle differing considerably from 0° or 90°) to the axis, the curves are approximately straight lines perpendicular to the principal section of the plate. Two such plates placed with their principal planes at right angles to one another give straight lines bisecting the angle between the principal planes. On this principle Savart constructed the polariscope which bears his name. It consists of two such plates and an analyser, and forms a very delicate test of the presence of polarisation. The lines are of course always in the direction described above, and the delicacy of the test increases in proportion as their direction becomes more and more nearly parallel or perpendicular to the original plane of vibration.

Bi-axal crystals exhibit a more complicated system of rings and crosses, or brushes, as they may in this case be better termed. If such a crystal be cut in a direction perpendicular to the line which bisects the angle between the two optic axes (or the middle line, as it is called), the extremity of each of the axes will be surrounded with rings similar to those described in the case of the uni-axal crystals. The larger rings, however, are not strictly circles, but are distorted and drawn out towards one another; those which are larger still meet at a point midway between the centres and form a figure of 8 or lemniscata; beyond this they form curves less and less compressed towards the crossing point, and approximate more and more nearly to an oval. (*See* Fig. 26; also Plate, Figs. 3 & 4.)

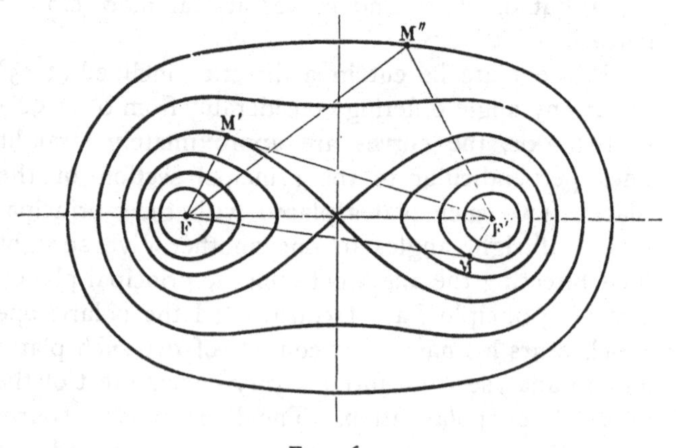

FIG. 26.

The vibrations of the two rays emerging from any point of a bi-axal crystal are as follows:—Of the two rays produced by the double refraction of a bi-axal crystal neither follows the ordinary law of refraction; but one does so more nearly than the other, and is on that account called for convenience the ordinary ray. And if through any point of the field of view we draw two lines to the points where the optic axes emerge, the directions of vibration of the two rays will be those of the bisectors of the angles made by the two lines. If, therefore, the crystal be so placed that the line joining the extremities of the two axes coincides with the plane of vibration of either polariser or analyser, it is not difficult to see that there will be a black cross passing through the centre of the field with one pair of arms in the line joining the extremities of the axes and the other pair at right angles to it. But if the plate be turned in its own plane round the central point, the points for which the vibrations are parallel or perpendicular to those of the polariser or analyser will no longer lie in straight lines passing through the centre, but will form two branches of a hyperbolic curve, passing respectively through the extremities of the optic axes.

If the analyser be turned round, the dark hyperbolic brushes, or the black cross, will undergo the changes analogous to those shown by the cross in the case of uni-axal crystals; but the most interesting effects are those seen when the polariser and analyser are crossed, and the crystal is turned in its own plane.

The angle between the optic axes in different kinds of crystals varies very much; in those where the angle is small it is easy to exhibit both at once in the field of view, but in others where the angle is large it is necessary to tilt the crystal so as to bring the two successively into view. In the latter case the crystal is sometimes cut in a direction perpendicular to one of the axes. The rings are then nearly circular, especially towards the centre, and in that respect they resemble those of a uni-axal crystal; but the character of the specimen can never be mistaken, because the rings are intersected by a black bar, or two arms in the same straight line, instead of by four arms at right angles to one another, as would have been the case if the crystal had been uni-axal. The following are the angles made by the optic axes in a few crystals :—

ORTHORHOMBIC CRYSTALS.

		Apparent Angle.	Actual Angle.
Aragonite . .	−	71° 31′	43° 32′
Cerceosite (Lead Carbonate) .	−	16° 54′	8° 7′
Chrysoberyll .	+	84° 43′	45° 20′
Chrysolite . .	+	—	88° 54′
Topaz . .	+	120° to 97°	65° to 55°
Anhydrite . .	+	71° 31′	43° 32′
Barytes . .	+	59° 6′	35° 4′
Potassio Sodium Tartrate (Rochelle Salt)	+	76° red 56° violet	
Ammonium Tartrate		59° 35′ (D)	38° 2′
Ammonio Sodium Tartrate		62° red 100° violet	46° 70°
Nitre . . .	−	9° 17′	6° 10′

OBLIQUE CRYSTALS.

Diopside . .	+	111° 34′	58° 59′
Epidote . .	−	115° 20′ to 92° 4′	87° 49′ to 74°

Oblique Crystals—*cont.*

		Apparent Angle.	Actual Angle.
Felspar (Orthoclase)	−	121° 6′ 20° 45′ (ignited)	69° 43′ 13° 34′
Sphene . .	+	56° to 53° 30′ red 34° to 32° 30′ blue	
Dichroite . .	−	Ceylon 125° 16′ Haddam 60° 46′	70° 23′ 37° 48′
Axinite . .	−	158°	71° 30′
Gypsum . .	+	max. varying with temperature.	75° 40′
Borax . .	+	59° 8′	39° 14′
Mica . . .	+	0° to 75°	

These angles are determined by placing the crystal for examination into an apparatus adapted to show the rings, and attaching it to an arm whereby the plate can be turned about an axis in its own plane. The axis is furnished with a circle divided into degrees and seconds, and an index. (*See* Fig. 27.) If this axis be horizontal, the plate is so placed that the line joining the centres of the two systems of rings is vertical, and the crystal is first turned so as to bring one centre into the centre of the field of view (usually marked by cross wires); the index is then read, and the crystal turned so as to bring the centre of the second system of rings to the centre of the field. The index is again read,

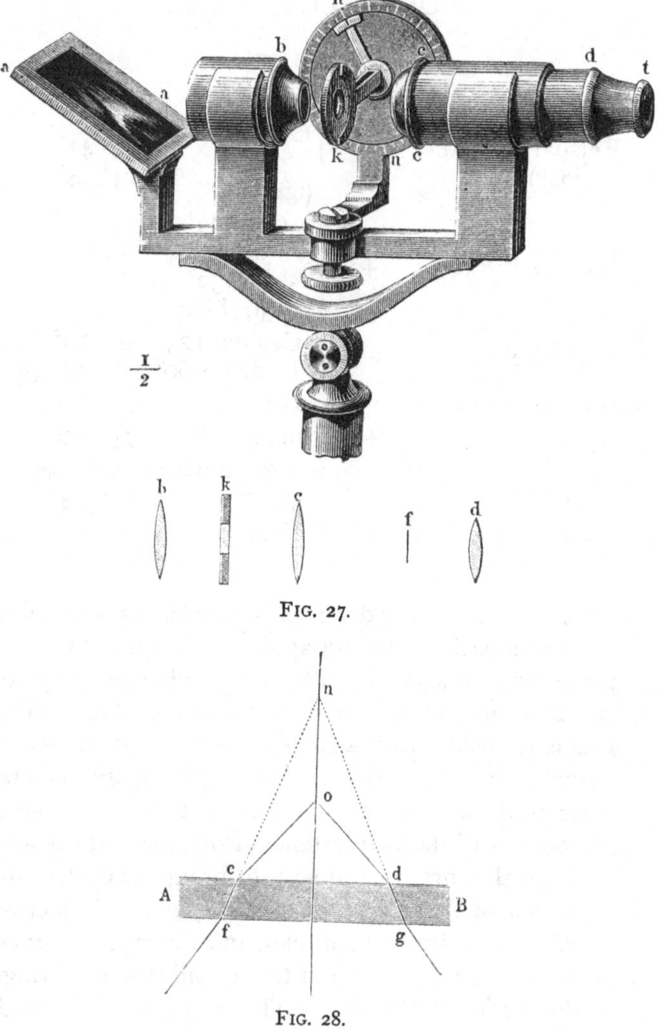

FIG. 27.

FIG. 28.

and the difference of the two readings noted. This, however, gives not the true angle of the optic axes, but the apparent angle in air—that is, the angle between the rays as affected by refraction on emerging from the crystal. (*See* Fig. 27.) In the annexed figure (28) A B represents the crystal plate, *o* the position of the eye, and *f c*, *g d* the directions of the axes in the crystal; the observed angle is *c o d*. But if the (mean) refractive index of the crystal is known, the true angle, *c n d*, may be calculated.

In bi-axal crystals the optic axes have, in the majority of cases, different angles for different colours, and where no limits are imposed on the direction of the first mean-line (the line bisecting the acute angle of the optic axes) by the conditions of symmetry of the crystallographic system, the mean-lines for different colours may be, and usually are, slightly inclined to each other, or, as it is termed, *dispersed*.

In the orthorhombic system the first mean-line for all colours coincides in direction with one of the crystallographic axes: the optic axes always lying in one or other of the axial planes (viz. in a plane containing the particular axis and one other crystallographic axis).

The optic axes for either the more or the less refrangible rays may be the more widely separated: it is rare for them not to present differences in this respect. The stauroscopic figures are orthosymmetrically divided by two lines passing through the centre of the field, one traversing the "eyes" of the optic axes, the other perpendicular to this line.

In the clinorhombic (or oblique) system:

1. The first mean-line may be parallel to the perpendicular axis. The planes containing the optic axes belonging to different colours will, in this case, all intersect in this line. The resulting stauroscopic figure presents symmetry to this axis, *i.e.* to the centre of the field, but is otherwise not symmetrical; and the angles of the optic axes may vary for every colour. Borax is an example of this.

2. The first mean-line for each colour may lie in the plane containing the oblique axes (the one plane of symmetry) of the system. The planes containing the optic axes (and, so, the second mean-lines) may lie in this plane. In this case the trace of this plane divides euthysymmetrically the stauroscopic figure, the mean-lines for different colours being dispersed in this plane. Where the optic axes for different colours have different degrees of divergence, the two "eyes" of the figure are very dissimilar, the one being larger than the other; where the divergence is not great, the colour that is outside in the one is inside in the other. Diopside, gypsum, and sphene are examples of this variety. (*See* Plate at end.)

3. And while the mean-lines lie in the plane of symmetry, the planes of the optic axes for different colours may be perpendicular to this plane (the second mean-line therefore being parallel to the perpendicular axis). In this case the stauroscopic figure is of course euthysymmetrical to the trace of the plane of symmetry which is now perpendicular to the lines joining the "eyes" of the figure. This kind of dispersion, then, shows the "eyes" for the red and blue, for example, dispersed on lines parallel to each other. Of this, adularia affords an example.

Lastly, in the anorthic system, the mean-lines for different colours may lie in different planes, as may also the optic axes which they bisect, and in consequence, as might be expected from the absence of any plane of symmetry in this system, the stauroscopic figures present no symmetry.

There are crystals of the orthorhombic and clinorhombic systems, moreover, which present a very remarkable variety of dispersion. Thus in brookite chrysoberyl, some of the alkali bibasic salts of tartaric acid, ammonium thallate, magnesio-ammonium, chromate, &c., the optic axes for the two extremities of the spectrum lie in planes at right angles to one another, both passing through the same middle line. If the systems of rings be examined with light which has been so widely dispersed that the portion illuminating the field in any given position is practically monochromatic, and the position of the instrument shifted through the different parts of the spectrum (or, what is more convenient, if the different parts of the spectrum be successively thrown on the polariscope by means of a totally reflecting prism), the optic axes will be seen to draw gradually together until the figure closely resembles that of a uni-axal crystal; after which the axes open out in a direction at right angles to the former, until they have attained their greatest expansion. This experiment requires a strong light, but it is instructive as showing the exact distribution of the optic axes for different rays.

In bi-axal crystals, probably in all cases, but notably in gypsum, the distribution of the optic axes varies with the temperature. When the crystal is heated the angle between the optic axes diminishes until the crystal appears uni-axal; with a further increase of temperature the axes again open out, but in a direction at right angles to the former. When the crystal is cooled, the axes generally resume their original directions. Sometimes, however, when the heating has been carried to a great degree, or has been continued for a long time, the axes never completely return to their normal condition; and in such a case the crystal may appear permanently uni-axal. Such an appearance, when permanent, has been considered a test of former heating; and this phenomenon, when presented by crystals found in a state of nature, may be taken as evidence that the rocks in which they have been formed have been subject to high temperatures.

In the production and examination of the rings hitherto described, we have used light which has been plane-polarised and plane-analysed; but there is nothing to prevent our polarising the light or analysing it circularly, or indeed doing both.

If a quarter-undulation plate be placed between the polariser and the crystal to be examined, with its axis inclined at 45° to the plane of original vibration, the light will fall upon the crystal in a state of circular polarisation; and as the polarisation will then have no reference to any particular plane of vibration, the black cross will disappear. A system of rings will be produced, but they will be discontinuous. At each quadrant, depending upon the position of the analyser, the rings will be broken, the portions in opposite quadrants being contracted or expanded, so that in passing from one quadrant to the next the colours pass into their complementaries. If either the direction of the axis of the quarter-undulation plate be changed from 45° on one side to 45° on the other side of the plane of vibration of the polariser, or if the crystal be changed for another of an opposite character

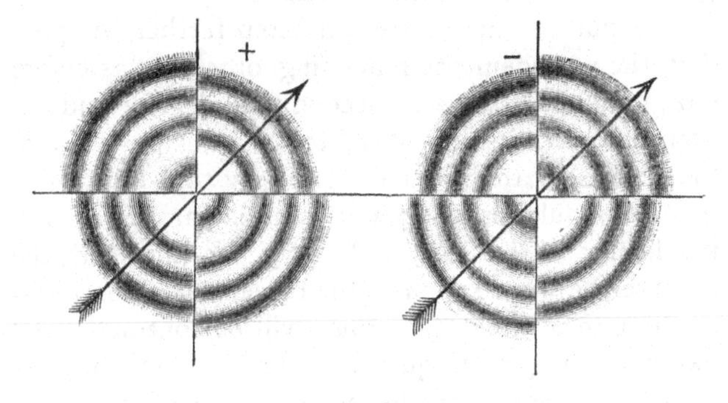

FIG. 29.

(*i.e.* negative for positive, or *vice versâ*), the quadrants which were first contracted will be expanded, and those which were first expanded will be contracted. Hence for a given position of the quarter-undulation plate the appearance of the rings will furnish a means of determining the character of the crystal under examination.

Similar effects are produced if the quarter-undulation plate be placed between the crystal and the analyser; that is, if the light be analysed circularly.

In the case of bi-axal crystals under the action of light polarised or analysed circularly, the black brushes are wanting, but they are replaced by lines of the same form, marking where the segments of the lemniscatas pass from given colours into their complementaries.

If the light be both polarised and analysed circularly, all trace of direction will have disappeared. In uni-axal crystals the rings will take the form of perfect circles without break of any kind ; and in bi-axal they will exhibit complete lemniscatas.

To pursue this matter one step further, suppose that, the arrangements remaining otherwise as before (viz., first, the polariser ; secondly, a quarter-undulation plate with its axis at 45° to the principal plane of the polariser ; thirdly, a uni-axal crystal ; fourthly, a quarter-undulation plate with its axis parallel or perpendicular to the first ; and, lastly, the analyser), the analyser be turned round ; then in any position intermediate to 0° and 90° the rings will be contracted and extended in opposite quadrants until at 45° they are divided by two diagonals, on each side of which the colours are complementary. Beyond 45° the rings begin to coalesce, until at 90° the four quadrants coincide again. During this movement the centre has changed from bright to dark. If the motion of the analyser be reversed, the quadrants which before contracted now expand, and *vice versâ*. Again, if the crystal be replaced by another of an opposite character—say positive for negative—the effect on the quadrants of the rings will be reversed. This method of examination, therefore, affords a test of the character of a crystal.

A similar process applies to bi-axal crystals ; but in this case the diagonals interrupting the rings are replaced by a pair of rectangular hyperbolas, on either side of which the rings expand or contract, and the effect is reversed by reversing the motion of the analyser, or by replacing a positive by a negative crystal. The test experiment may then be made by turning the analyser slightly to the right or left, and observing whether the rings appear to advance to, or recede from, one another in the centre of the field. In particular if, the polariser and analyser being parallel, the first plate have its axis in a N.E. direction to a person looking through the analyser, the second plate with its axis at right angles to the former, and the crystal be so placed that the line joining the optic axes by N.S., then on turning the analyser to the right the rings will advance towards one another if the crystal be negative, and recede if it be positive.

CHAPTER IX.

COMPOSITION OF COLOURS BY POLARISED LIGHT.[1]

THE results of combining two or more colours of the spectrum have been studied by Helmholtz, Clerk Maxwell, Lord Rayleigh, and others. And the combinations have been effected sometimes by causing two spectra at right angles to one another to overlap, and sometimes by bringing images of various parts of a spectrum simultaneously upon the retina. Latterly also W. von Bezold has successfully applied the method of binocular combination to the same problem (*Poggendorff*, Jubelband, p. 585). Some effects, approximating more or less to these, may be produced by chromatic polarisation.

Complementary Colours.—First, as regards complementary colours. If we use a Nicol's prism N as polariser, a plate of quartz Q cut perpendicularly to the axis, and a double-image prism P as analyser, we shall, as is well known, obtain two images whose colours are complementary. If we analyse these images with a

[1] Extracted from the Proceedings of the Royal Society, No. 153, 1874. See also Proceedings of the Royal Institution, 1874.

prism, we shall find, when the quartz is of suitable thickness, that each spectrum contains a dark band, indicating the extinction of a certain narrow portion of its length. These bands will simultaneously shift their position when the Nicol N is turned round. Now, since the colours remaining in each spectrum are complementary to those in the other, and the portion of the spectrum extinguished in each is complementary to that which remains, it follows that the portion extinguished in one spectrum is complementary to that extinguished in the other. And in order to determine what portion of the spectrum is complementary to the portion suppressed by a band in any position we please, we have only to turn the Nicol N until the band in one spectrum occupies the position in question, and then to observe the position of the band in the other spectrum. The combinations considered in former experiments are those of simple colours; the present combinations are those of mixed tints, viz. of the parts of the spectrum suppressed in the bands. But the mixture consists of a prevailing colour corresponding to the centre of the band, together with a slight admixture of the spectral colours immediately adjacent to it on each side.

The following results given by Helmholtz may be approximately verified :—

Complementary Colours.

Red.	Orange.
Yellow.	Yellow-green
Green-blue.	Cyanic-blue.
Indigo-blue.	Violet.

When in one spectrum the band enters the green, in the other a band will be seen on the outer margin of the red, and a second at the opposite end of the violet; showing that to the green there does not correspond one complementary colour, but a mixture of violet and red, *i.e.* a reddish purple.

Combination of two Colours.—Next, as to the combination of two parts of the spectrum, or of the tints which represent those parts. If, in addition to the apparatus described above, we use a second quartz plate Q_1 and a second double-image prism P_1, we shall form four images—say O O, O E, E O, E E. And if A, A′ be the complementary tints extinguished by the first combination Q P alone, and B, B′ those extinguished by the second Q_1 P_1 alone, then it will be found that the following pairs of tints are extinguished in the various images.

Image.	Tints extinguished.
O O	B, A
O E	B′, A′
E O	B′, A
E E	B, A′

It is to be noticed that in the image O E the combination Q_1 P_1 has extinguished the tint B′ instead of B, because the vibrations in the image E were perpendicular to those in the image O formed by the combination Q P. A similar remark applies to the image E E.

The total number of tints which can be produced

by this double combination Q P, Q_1 P_1 is as follows :—

4 single images
6 overlaps of two
4 overlaps of three
1 overlap of four

Total 15

Collateral Combinations.—The tints extinguished in the overlap O O + E O will be B, A, B′, A; but since B and B′ are complementary, their suppression will not affect the resulting tint except as to intensity, and the overlap will be effectively deprived of A alone; in other words, it will be of the same tint as the image O would be if the combination Q_1 P_1 were removed. Similarly the overlap O E + E E will be deprived effectually of A′ alone; in other words, it will be of the same tint as E, if Q_1 P_1 were removed. If, therefore, the Nicol N be turned round, these two overlaps will behave in respect of colour exactly as did the images O and E, when Q P was alone used. We may, in fact, form a table thus :—

Image.	Colours extinguished.
O O + E O	$B + A + B' + A = B + B' + A = A$
O E + E E	$B' + A' + B + A' = B + B' + A' = A'$

And since the tints B, B′ have disappeared from each of these formulæ, it follows that the second analyser P may be turned round in any direction without altering the tints of the overlaps in question.

In like manner we may form the table—

OO + EE	$B + A + B + A' = B + A + A' = B$
OE + EO	$B' + A' + B' + A = B' + A + A' = B'$

Hence if the Nicol N be turned round, these overlaps will retain their tints; while if the analyser P_1 be turned, their tints will vary, although always remaining complementary to one another.

There remains the other pair of overlaps, viz.:—

OO + OE	$B + A + B' + A$
EO + EE	$B' + A + B + A'$

Each of these is deprived of the pair of complementaries A, A', B, B'; and therefore each, as it would seem, ought to appear white of low illumination, *i.e.* grey. This effect is, however, partially masked by the fact that the dark bands are not sharply defined like the Fraunhofer lines, but have a core of minimum or zero illumination, and are shaded off gradually on either side until a short distance from the core the colours appear in their full intensity. Suppose, for instance, that B' and A' were bright tints, the tints resulting from their suppression would be bright. On the other hand, the complementary tints A and B would be generally dim, and the image $B + A$ bright, and the overlap $B + A + B' + A$ would have as its predominating tint that of $B + A$. And similarly in other cases.

There are two cases worth remarking in detail, viz., first, that in which

$$B = A', \ B' = A$$

i.e. when the same tints are extinguished by the combination Q P and by Q_1 P_1. This may be verified by either using two similar quartz plates Q, Q_1, or by so turning the prism P_1 that the combination Q_1 P_1 used alone shall give the same complementary tints as Q P when used alone. In this case the images have for their formulæ the following :—

O O	O E	E O	E E
$A + A'$	$A + A'$	$2A$	$2A'$

in other words, O O and E O will show similar tints, and E O, E E complementary. A similar result will ensue if $B = A$, $B' = A'$.

Again, even when neither of the foregoing conditions are fulfilled, we may still, owing to the breadth of the interference bands, have such an effect produced that sensibly to the eye

$$B + A = B' + A''$$

and in that case

$$\begin{aligned} B' + A &= B + A - A' + A \\ &= B + A' + 2A - 2A' \end{aligned}$$

which imply that the images O O and O E may have the same tint; but that E O and E E need not on that account be complementary. They will differ in tint in this, that E E, having lost the same tints as E O, will have lost also the tint A, and will have received besides the addition of two measures of the tint A′.

Effect of Combination of two Colours.—A similar

train of reasoning might be applied to the triple overlaps. But the main interest of these parts of the figure consists in this, that each of the triple overlaps is complementary to the fourth single image; since the re-combination of all four must reproduce white light. Hence the tint of each triple overlap is the same to the eye as the mixture of the two tints suppressed in the remaining image. And since by suitably turning the Nicol N or the prism P_1, or both, we can give any required position to the two bands of extinction, we have the means of exhibiting to the eye the results of the mixture of tints due to any two bands at pleasure.

Effect of Combinations of three Colours.—A further step may be made in the combination of colours by using a third quartz Q_2 and a third double-image prism P_2, which will give rise to eight images. And if C C′ be the complementaries extinguished by the combination Q_2 P_2, the formulæ for the eight images may be thus written:—

O O O	$C + B + A$
O O E	$C + B' + A'$
O E O	$C' + B' + A$
O E E	$C' + B + A'$
E O O	$C' + B + A$
E O E	$C' + B' + A'$
E E O	$C + B' + A$
E E E	$C + B + A'$

The total number of combinations of tint given by the compartments of the complete figure will be—

$\frac{8}{1}$ = 8 single images.

$\frac{8.7}{1.2}$ = 28 overlaps of two.

$\frac{8.7.6}{1.2.3}$ = 56 " " three.

$\frac{8.7.6.5}{1.2.3.4}$ = 70 " " four.

$\frac{8.7.6}{1.2.3}$ = 56 " " five.

$\frac{8.7}{1.2}$ = 28 " " six.

$\frac{8}{1}$ = 8 " " seven.

1 = 1 " " eight.

Total . . 255

The most interesting features of the figure consists in this, that the subjoined pairs are complementary to one another, viz. :—

OOO C+B+A	EOE C′+B′+A′
EOO C′+B+A	OOE C+B′+A′
EEO C+B′+A	OEE C′+B+A′
EEE C+B+A′	OEO C′+B′+A

And if the prisms P, P_1, P_2 are so arranged that the separations due to them respectively are directed parallel to the sides of an equilateral triangle, the images will be disposed thus :—

	O E O	O O O	
E E O	E O O	O E E	O O E
	E E E	E O E	

The complementary pairs can then be read off, two horizontally and two vertically, by taking alternate pairs, one in each of the two vertical, and two in the one horizontal row. And each image will then represent the mixture of the three tints suppressed in the complementary image.

Low-tint Colours.—A slight modification of the arrangements above described furnishes an illustration of the conclusions stated by Helmholtz, viz. that the low-tint colours (*couleurs dégradées*), such as russet, brown, olive-green, peacock-blue, &c., are the result of relatively low illumination. He mentioned that he obtained these effects by diminishing the intensity of the light in the colours to be examined, and by at the same time maintaining a brilliantly illuminated patch in an adjoining part of the field of view. If, therefore, we use the combination N, Q, P, P_1 (*i.e.* if we remove the second quartz plate), we can, by turning the prism round, diminish to any required extent the intensity of the light in one pair of the complementary images and at the same time increase that in the other pair. This is equivalent to the conditions of Helmholtz's experiments; and the tints in question will be found to be produced.

INDEX.

Spottiswoode & Co., Printers, New-street Square, London.

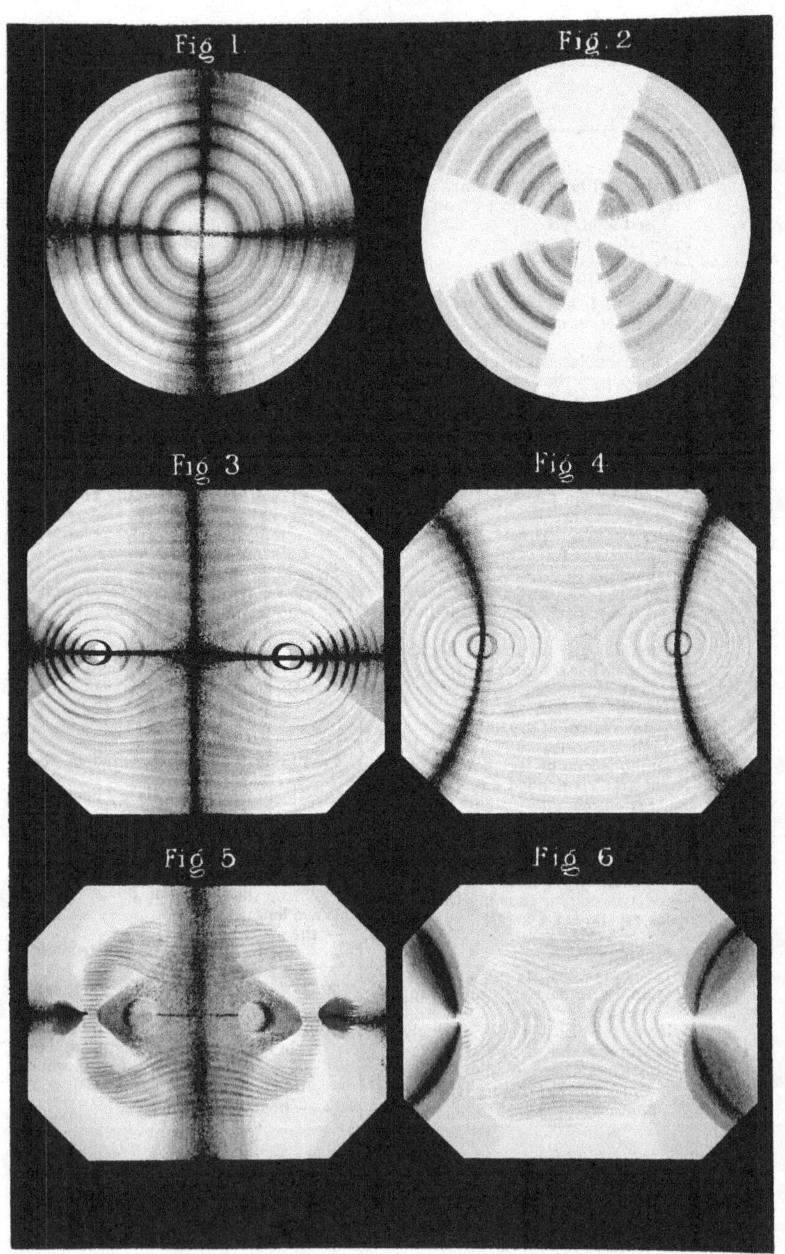

HANHART, CHROMO-LITH.

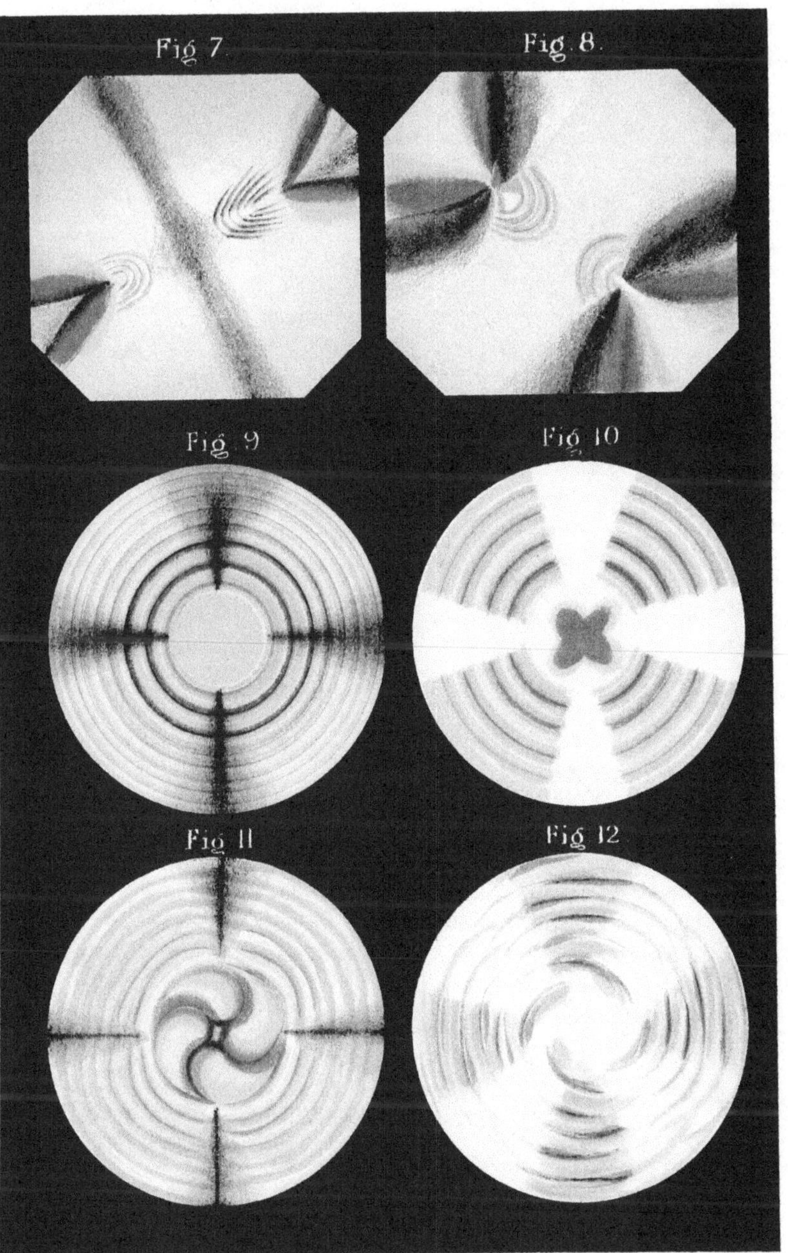

HANHART, CHROMO LITH

Printed in the United States
By Bookmasters